Hermann Flohn (Ed.)
TROPICAL RAINFALL ANOMALIES AND CLIMATIC CHANGE

Geographisches Institut
der Universität Kiel
Neue Universität

BONNER METEOROLOGISCHE ABHANDLUNGEN

Heft 31 (1984) (ISSN 0006-7156)
Herausgeber: MICHAEL HANTEL

Hermann Flohn (Ed.)

TROPICAL RAINFALL ANOMALIES AND CLIMATIC CHANGE

Dümmlerbuch 7591

DÜMMLER · BONN

Address of the authors: Anschrift der Verfasser:

                          c/o: Prof. Dr. H. Flohn
                                  Meteorologisches Institut
                                  der Universität
                                  Auf dem Hügel 20
                                  D-5300 Bonn 1

ISBN 3-427-75911-3

Alle Rechte, insbesondere auch die der Übersetzung, des Nachdrucks, des Vortrages, der Verfilmung und Radiosendungen sowie jeder Art der fotomechanischen Wiedergabe und der Speicherung in Datenverarbeitungsanlagen, auch auszugsweise, vorbehalten. Das Fotokopieren einzelner Seiten ist nicht gestattet, mit Ausnahme der in §§ 53, 54 UrhG ausdrücklich genannten Sonderfälle.

© 1984 Ferd. Dümmlers Verlag, Kaiserstraße 31-37 (Dümmlerhaus), 5300 Bonn 1

Printed in Germany by Richard Schwarzbold, 5305 Witterschlick bei Bonn

Hermann Flohn (Guest Editor)
GENERAL INTRODUCTION

During the 1960's and 1970's efforts were made at the Meteorological Institute of Bonn University to assemble a data bank of sufficiently long monthly rainfall series from tropical and subtropical latitudes. It was intended to investigate the teleconnections between rainfall anomalies in low latitudes and circulation anomalies in middle and higher latitudes. One of the most remarkable anomaly types is correlated with the El Niño phenomenon and similar but weaker features in the Atlantic area. In the course of the years extended and coherent station series were compiled; further, they were combined to representative area averages. Selected results were published in the *Bonner Meteorologische Abhandlungen* (BMA Nos. 1, 7, 8, 9, 11, 26). The present issue is a continuation of this effort.

In the first paper H. Behrend deals with the rainfall anomalies during El Niño events, using data along the equator from East Africa across the Indian and Pacific Oceans to the Caribbean.

The second paper gives a summary of the diploma thesis of C. Becker which is based on long station records from northern South America; Becker's data for Curaçao are included in Behrend's study as one of the reference stations. The unpublished results of B. Schweitzer and H. Raatz on India and Sri Lanka have been reviewed by H. Fleer (*Mausam*, 35, 1984, 135-144).

In the third paper we investigate the incomplete records of Nauru for the years before 1914. These data are shown to be already useful for an investigation of the remarkable correlation between rainfall and the zonal wind component: rain frequency is maximum for westerly winds and minimum for easterly winds. This correlation has been reported years ago (Flohn, 1957) using a comprehensive data set of the Atlantic. It seems to be a quite general but hardly understood phenomenon in the equatorial zone.

The fourth paper (H.-P. Junk) deals with the complete data set of Nauru. The long record of this station has been once more carefully reconstructed by filling its gaps with significantly correlated "neighbouring" stations; this reconstruction deviates in minor details from those used by Fleer (BMA 26) and Behrend.

Among the most remarkable features of air-sea interaction in equatorial latitudes are the exchange processes of CO_2 and H_2O strongly varying between "normal" upwelling and El Niño (downwelling). The fifth paper of this issue (of K.-H. Weber) gives some statistical background; its main purpose is a more detailed discussion of its possible role for the composition of the atmosphere and for global climatic fluctuations on time-scales between 1 and 1000 years. Due to the multi-disciplinary nature of these exchange processes, the considerations here remain partly speculative, but it is hoped that they may serve as a useful starting point for further discussion.

Hermann Flohn (Herausgeber dieses Heftes)
ALLGEMEINE EINFÜHRUNG

In den 60er und 70er Jahren wurde am Meteorologischen Institut der Universität Bonn damit begonnen, eine Datenbank für hinreichend lange Zeitreihen des Monatsniederschlages in tropischen und subtropischen Breiten anzulegen. Die Absicht war, Telekonnektionen zwischen Niederschlagsanomalien in niedrigen Breiten und Zirkulationsanomalien in mittleren und höheren Breiten zu untersuchen. Einer der bemerkenswertesten Anomalietypen ist korreliert mit dem El Niño-Phänomen und ähnlichen, wenn auch schwächeren, Vorgängen im Bereich des Atlantik. Im Lauf der Jahre wurden ausgedehnte und zusammenhängende Stationsreihen bereitgestellt; sie wurden darüber hinaus zu repräsentativen Flächenmitteln kombiniert. Ausgewählte Ergebnisse wurden in der Reihe *Bonner Meteorologische Abhandlungen* veröffentlicht (BMA Nrn. 1, 7, 8, 9, 11, 26). Das vorliegende Heft ist eine Fortsetzung dieser Bemühungen.

Im ersten Beitrag befaßt sich H. Behrend mit den Niederschlagsanomalien während eines El Niño-Ereignisses; er verwendet Daten am Äquator von Ostafrika über den Indischen und Pazifischen Ozean bis hin zur Karibik.

Der zweite Artikel gibt eine Zusammenfassung der Diplomarbeit von C. Becker, die auf langen Stationsreihen aus dem nördlichen Südamerika beruht; Becker's Daten für Curacao sind in Behrend's Arbeit als eine der Referenzstationen enthalten. Die unveröffentlichten Resultate von B. Schweitzer und H. Raatz über Indien und Sri Lanka sind von H. Fleer (*Mausam*, 35, 1984, 135-144) besprochen worden.

Im dritten Beitrag untersuchen wir die unvollständige Reihe von Nauru für die Jahre vor 1914. Diese Daten sind bereits recht wertvoll für eine Untersuchung der bemerkenswerten Korrelation zwischen dem Regen und der zonalen Windkomponente: Die Regenhäufigkeit ist maximal für Westwind und minimal für Ostwind. Diese Korrelation wurde bereits vor Jahren auf Grundlage eines umfassenden Datensatzes für den Atlantik beschrieben (Flohn, 1957). Sie scheint ein ganz allgemeines aber bisher kaum verstandenes Phänomen der Äquatorialzone zu sein.

Der vierte Artikel (H.-P. Junk) befaßt sich mit dem vollständigen Datensatz von Nauru. Die lange Reihe dieser Station wurde erneut sorgfältig rekonstruiert durch Füllen der Lücken mit signifikant korrelierten "Nachbarstationen"; diese Rekonstruktion weicht in kleineren Einzelheiten von denen von Fleer (BMA 26) und Behrend ab.

Zu den auffälligsten Erscheinungen der atmosphärischen ozeanischen Wechselwirkung in äquatorialen Breiten gehören die Austauschprozesse von CO_2 und H_2O, die starke Unterschiede zwischen "normalem" Aufquellen und El Niño (Absinken) zeigen. Der fünfte Beitrag dieses Heftes (von K.-H. Weber) gibt statistisches Grundlagenmaterial; sein Hauptzweck ist eine eingehende Diskussion der möglichen Rolle dieses Prozesses für die Zusammensetzung der Atmosphäre und für globale Klimaschwankungen auf Zeitskalen von 1 bis 1000 Jahren. Wegen der fachübergreifenden Natur dieser Austauschvorgänge sind die hier gegebenen Betrachtungen teilweise spekulativ; es ist zu hoffen, daß sie als nützlicher Ausgangspunkt für weitere Diskussionen dienen können.

ACKNOWLEDGMENTS

Part of this study has been funded by the "Rheinisch-Westfälische Akademie der Wissenschaften", Düsseldorf, and the "Kommission der Europäischen Gemeinschaften", Brussels. The drawings have partly been carried out by Mr. N. Wilinski. Mrs. C. Frese took care of the manuscript.

TABLE OF CONTENTS - INHALT

Page

H. Flohn (Guest Editor):

GENERAL INTRODUCTION VII

A) H. Behrend:

TELECONNECTIONS OF TROPICAL RAINFALL ANOMALIES 1
AND THE SOUTHERN OSCILLATION

Abstract 2

A.1　Introduction 3

A.2　Data 6
　　　a) Time series of rainfall 6
　　　b) Time series of other meteorological elements 11
　　　c) Properties of selected time series 14

A.3　Autocorrelation analysis 15

A.4　Cross-correlation analysis 19
　　　a) Calculations 19
　　　b) Results 21
　　　c) Discussion 28

A.5　Composite analysis of El Niño rainfall anomalies 36

A.6　Concluding comments 44

Appendix: Significant cross-correlations 47

B) C. Becker and H. Flohn:

TELECONNECTIONS IN THE CARIBBEAN AND 51
NORTHERN SOUTH AMERICA AND
THE SOUTHERN OSCILLATION

C) H. Flohn:

ZONAL SURFACE WINDS AND RAINFALL IN 57
THE EQUATORIAL PACIFIC AND ATLANTIC

D) H.-P. Junk:

NAURU RAINFALL 1893 - 1977; 67
A STANDARD COMPOSITE RECORD

	Page
E) K.-H. Weber and H. Flohn:	
OCEANIC UPWELLING AND AIR-SEA-EXCHANGE OF CARBON DIOXIDE AND WATER VAPOR AS A KEY FOR LARGE-SCALE CLIMATIC CHANGE?	73
Abstract	74
E.1 Introduction	75
E.2 SST-CO_2 Correlations	76
E.3 SST-H_2O Correlations	80
E.4 Parallel variations of atmospheric CO_2 and H_2O	85
E.5 Abrupt climatic changes under natural conditions: the role of CO_2	86
E.6 A geophysical interpretation of large-scale climatic changes on the 10^2-10^3-year scale	93
REFERENCES	99

A   Hartmut Behrend:

TELECONNECTIONS OF
TROPICAL RAINFALL ANOMALIES
AND THE
SOUTHERN OSCILLATION

ABSTRACT

Based on long-term monthly rainfall series from a selection of tropical area-averages and individual stations, spatial cross-correlations with different lags have been calculated for an equatorial area extending from East Africa across the Indian Ocean and the Pacific to the Caribbean, i.e. over about $250°$ longitude. For a more detailed study of the time relations, systematic autocorrelation studies have been made; a tendency for persistence is frequently limited to some seasons. In comparison to the results of Rasmussen and Carpenter (1982) - based mainly on surface winds and sea-surface temperatures - a composite analysis of the rainfall anomalies depending on longitude and time (during a 3 year El Niño-epoch) is given.

A remarkable month-to-month persistence is found for the greatest part of the Indian Summer Monsoon (July (7) - October (10)), the Indonesian SE-Monsoon (7 - 12), for the greatest part of the year (6 - 4) in the equatorial Central Pacific, for the rainy seasons at Galapagos (2 - 8) and Curaçao (10 - 2) and the little rainy season of the East African coast (9 - 12). The Southern Oscillation Index (Wright) is positively correlated with rainfall of the SW-monsoon of India, of the SE-monsoon of Indonesia, of the rainy season at Curaçao and with the following winter rains of Hawaii; it is negatively correlated with the rainfall anomalies of the equatorial Pacific islands and with the second intermonsoonal season (10 - 11) at Sri Lanka. Similar teleconnections are found between the equivalent rainfall anomalies. The rainfall anomalies during an El Niño event fit well into this scheme of teleconnections. The peak of the anomalies moves along the Pacific equator with a speed of about 2000 km/month towards E; in a poleward direction the speed is limited to about 100 km/month.

Most of these teleconnections occur during the second half of the year, when the southern hemispheric circulation (and the SE-trades) reach their peak, controlling the climates of the corresponding areas even beyond the equator.

A.1 INTRODUCTION

Dramatic climatic anomalies, as in the El Niño-events of 1972 (Wooster and Guillan, 1974, Ramage, 1975), have demonstrated, through a series of simultaneous anomalies in far distant areas, the important role of large-scale tropical rainfall teleconnections. During 1972 heavy rainfall extended from the eastern Pacific up to the Philippines, while rainfall in Indonesia was reduced and India and the whole North African Sahel belt suffered simultaneously one of the heaviest droughts of this century (Fleer, 1975).

During the occurrence of an El Niño-event the SO[*] falls into its negative phase (Berlage, 1957, Troup, 1965, Wright, 1977) and the zonal Walker circulation in the Pacific (Bjerknes, 1969) is weak. Along the equator three well-defined Walker cells are stated: the largest one between Indonesia and the eastern Pacific, another between the interior of Brazil and the eastern Atlantic, while the third one between Indonesia and the eastern coast of Africa runs in a reverted sense, only during northern summer (Flohn, 1971). As the most important heat source - releasing latent heat via its enormous precipitations - Indonesia drives two of these Walker cells (Ramage, 1968). Hence, much attention is devoted to the RA's of Indonesia.

The primary source of tropical rainfall is the ITCZ. Therefore the variations of its location and intensity are of great importance for the teleconnections of tropical RA's. The moisture source of the ITCZ is the low-level convergence of the trade winds. The ITCZ is subject to a large seasonal march between about $25°N$ and $15°S$ in the monsoon regions of Africa and Asia, while this march is much weaker in the central and eastern Pacific and Atlantic (Fig. 1) due to the heat storage capa-

---

[*]
Acronyms used in this paper:
- acf — autocorrelation coefficient
- ccf — correlation coefficient
- ITCZ — Intertropical Convergence Zone
- NECC — North Equatorial Counter Current
- RA — Rainfall Anomaly
- SEC — South Equatorial Current
- SO — Southern Oscillation
- SOI — Southern Oscillation Index
- SPCZ — South Pacific Convergence Zone
- SST — Sea Surface Temperature

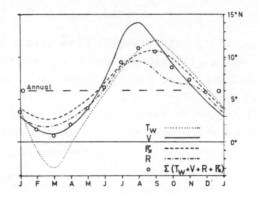

Fig. 1a)
 Latitude of meteorological equator at 25°W (Flohn, 1967).
 T_W = maximum of sea-surface temperature
 V = sign change of meridional wind component
 R̃ = maximum of thunderstorm frequency
 R = maximum of precipitation frequency

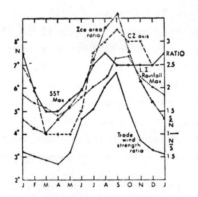

Fig. 1b)
 Latitudes of surface wind convergence zone axis (CZ axis), Line Islands rainfall maximum, sea-surface temperature maximum near 159°W (SST Max) as related to ratio of North Pacific trade wind strength to South Pacific trade wind strength and to ratio of Arctic ice area to Antarctic ice area. S/N denotes south greater, N/S, north greater. (Reproduced from Ramage et al., 1981.)

city of oceans. Here the ITCZ remains, during the whole year, north of the equator, partly because of the cold upwelling water at the equator and south of it (Flohn, 1967). During an El Niño the Pacific ITCZ moves closer to the equator than usual and the SPCZ moves to the northeast (Trenberth, 1976). This leads to heavy rainfall in the vicinity of the dateline, where both convergence zones are nearly merging. The atmospheric "center of action" moves from Indonesia into this region (Ichiye and Pettersen, 1963; Ramage, 1975). Heavy rainfall occurs over the central and eastern Pacific, while the Indian southwest monsoon rains are weak (Angell, 1981). The southwest Pacific (Donguy and Henin, 1980) and Indonesia are also dry (Nicholls, 1981). The SEC of the Pacific weakens in response to the weaker trades while the NECC strengthens and transports warm water from the western Pacific towards the east, intruding the cold equatorial eastern Pacific (Wyrtki, 1973, 1975; Barnett, 1977).

Many theories exist about the development of El Niño. Bjerknes (1969) pointed out, that the slowing down of the southeast trades in the Pacific should be the main precursor leading to a reduction of upwelling cold water and therefore to higher SST's. The atmospheric response to a positive SST-anomaly in the central and eastern Pacific has been investigated in several models. In a hemispheric model Rowntree (1972) shows the slowing down of the Walker circulation within the area of positive SST, while precipitation was enhanced to the west. Julian and Chervin (1978) studied the same topics in a global model. Wells (1979) was able to show in a hemispheric coupled ocean-atmosphere model the movement of a positive SST-anomaly from the east to the west in the equatorial Pacific, but his result of a velocity of 20° longitude within 80 days is unrealistic, because the model neglects advection. Wyrtki emphasized the oceanic response to an atmospheric forcing of the southeast trades upon the Pacific as the most important element for the development of an El Niño (Wyrtki, 1975, 1979). He suggested that after the relaxation of the trades an equatorial Kelvin wave travels from the West Pacific to the South American coast lowering the thermocline. He also drew attention to the role of the NECC in transporting warm water from the West Pacific (Wyrtki, 1973, 1975).

The purpose of this study is to contribute to the investigation of these teleconnections and their time-lags. Section 3 presents an autocorrelation analysis to identify the seasons of highest persistence of the anomalies of our time series. Section 4 presents a cross-correlation analysis describing the teleconnections, their intensity, seasonal dependence and sign. These fit rather well into the scheme of the SO. In section 5 the amplitude of the RA's during an average El Niño will be demonstrated with composites of selected time series.

During the work on this paper, the Super-Niño of 1982/83 developed, in many regards different from its medium-scale or weak predecessors. We refrain from discussing this event in the present paper.

A.2 DATA

a) Time series of rainfall

The rainfall series used in the correlation analysis are listed in Table 1; Fig. 2 shows their geographical position. Most of them are area-averages over larger areas. In this case only series with significantly positive correlations to several reference stations have been used. The advantages of such spatially coherent area-averages are:
a) the amount of data is drastically reduced;
b) small-scale effects of single rainfall stations, induced, e.g., by orography or inhomogeneities of the records, are minimized;
c) variations of the convective activity over larger areas represent large-scale fluctuations in the release of latent heat which affect atmospheric circulations and teleconnections crucially.

Table 1 includes the sources of the rainfall series. If the source is denoted by an asterisk, this area average has been evaluated by the present author (e.g., within the Indonesian Archipelago). For this purpose the rainfall stations (Table 2) were cross-correlated with nine (underlined) basic series. These basic series are quite distant from each other and represent contrasting climates. After removal of the seasonal variations simultaneous ccf's of the monthly RA's have been calculated, separately for each month, using the standard definitions.

Next the 12 monthly ccf's for each pair were averaged and mapped. Area-averages of rainfall were only used when in a specific region the average ccf with the basic series exceeded the 90 percent confidence limit by applying Student's t-test. With about 80 degrees of freedom this corresponds to a ccf of about 0.2. After this procedure the following area-averages were established: West Indonesia with Singapore, Madan, Pangalpinang and Pontianak; Java with Djakarta, Tjaltjap, Christmas Island, Pasuruan and Sumenep; Southeast New Guinea with Port

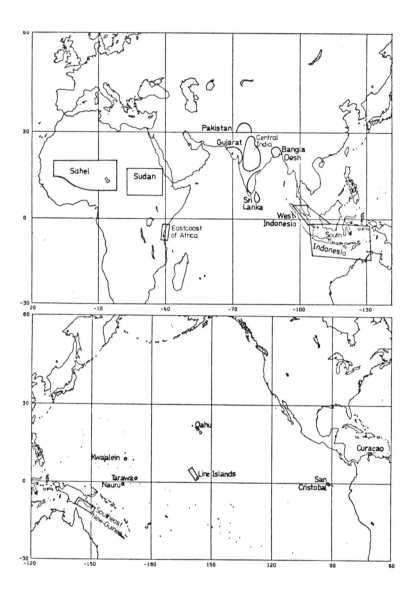

Fig. 2

Geographical positions of rainfall series used in cross-correlation analysis.

Table 1

Rainfall series used in cross-correlation analysis

Rainfall series	Time range	Source	Area/km^2	Number of stations
North Pakistan	1900-1960	Schweitzer (1978)	60 000	11
Gujarat	1875-1974	Schweitzer (1978)	150 000	27
Bangla Desh	1900-1960	Schweitzer (1978)	90 000	9
Central India	1870-1974	Schweitzer (1978)	450 000	10
Southeast India	1835-1974	Schweitzer (1978)	200 000	7
Sri Lanka	1875-1975	Raatz (1977)	60 000	13
West Indonesia	1879-1970	*	900 000	4
South Indonesia	1879-1970	*	3 300 000	13
Southeast New Guinea	1904-1972	*	500 000	6
Kwajalein	1945-1980	Taylor, MCD (1973)	-	1
Nauru	1892-1972	Fleer (1975)	-	1
Tarawa	1926-1980	Taylor, MCD (1973)	-	1
Line Islands	1910-1975	Meisner (1976)	200 000	3
Oahu	1891-1975	Meisner (1976)	5 000	9
San Cristobal	1950-1980	Becker (1982)	-	1
Curaçao	1895-1970	Becker (1982)	-	1
Sahel	1905-1973	Helbig (1976)	3 300 000	20
Sudan	1902-1973	Helbig (1976)	1 400 000	8
East coast of Africa	1893-1971	Helbig (1976)	200 000	6

MCD = Monthly Climatic Data for the World
* = area-average compiled in present study.

Table 1a

Position of rainfall stations

station	position	height
Kwajalein	8°43'N / 167°44'E	
Nauru	0°34'S / 166°55'E	3 m
Tarawa	1°21'N / 172°56'E	4 m
San Cristobal	1°09'S / 89°06'W	6 m
Curaçao	12°12'N / 68°58'W	8 m

Table 2
Rainfall series on the Indonesian Archipelago

	Station	Position	Height	Time range
	Kutaraja	5°31'N / 95°25'E	21 m	1879-1960
	Medan	3°14'N / 98°41'E	25 m	1879-1970
	Singapore	1°20'N / 103°50'E	18 m	1911-1960
	Padang	0°53'S / 100°21'E	3 m	1879-1960
S	Pangalpinang	2°10'S / 106°03'E	33 m	1888-1970
	Pontianak	0°01'S / 109°20'E	3 m	1879-1970
	Sandakan	5°52'N / 118°04'E	14 m	1879-1960
	Tarakan	3°20'N / 117°34'E	5 m	1911-1971
	Manokwari	0°53'S / 134°05'E	3 m	1901-1960
S	Kendari	4°36'S / 122°26'E	50 m	1908-1970
	Palu	0°41'S / 119°44 E	6 m	1908-1970
	Manado	1°30'N / 124°50 E	4 m	1879-1970
S	Makassar	5°04'S / 119°33 E	14 m	1879-1970
S	Ambon	3°42'S / 128°05 E	12 m	1879-1970
S	Djakarta	6°11'S / 106°50 E	6 m	1864-1972
S	Christmas Island	10°25'S / 105°43'E		1901-1972
S	Tjaltjap	7°44'S / 109°01'E	6 m	1879-1971
S	Pasuruan	7°38'S / 112°55'E	5	1888-1971
S	Sumenep	7°01'S / 113°51'E		1879-1970
S	Ampenan	3°43'S / 116°04'E	3 m	1895-1970
S	Waingapu	9°40'S / 120°20'E	12 m	1907-1970
S	Kupang	10°10'S / 123°34'E	15 m	1879-1970
S	Darwin	12°30'S / 130°54'E	29 m	1879-1975
	Daru	9°04'S / 143°12'E	5 m	1894-1972
	Kikori	7°25'S / 144°14'E	74 m	1913-1972
	Kerema	7°58'S / 145°45'E	6 m	1909-1972
	Port Moresby	9°30'S / 147°12'E	30 m	1891-1975
	Samarai	10°37'S / 150°39'E	41 m	1904-1972
	Kokoda	8°53'S / 147°44'E	366 m	1908-1972
	Lorengau	2°01'S / 147°19'E	16 m	1916-1972
	Kavieng	2°35'S / 150°48'E	7 m	1916-1972
	Rabaul	4°12'S / 152°11'E		1912-1972

The rainfall series were obtained by Prof. Dr. H. Flohn in 1972 from the Meteorological Service of Indonesia (Djakarta). Exceptions are Darwin, which was delivered in 1977 by the Australian Meteorological Bureau (Melbourne), and Singapore, Sandakan and Manokwari, which are taken from a tape kindly supplied by the National Center for Atmospheric Research (Boulder, Colorado).
S: Station used for area-average of South Indonesia
Underlined stations: Reference stations

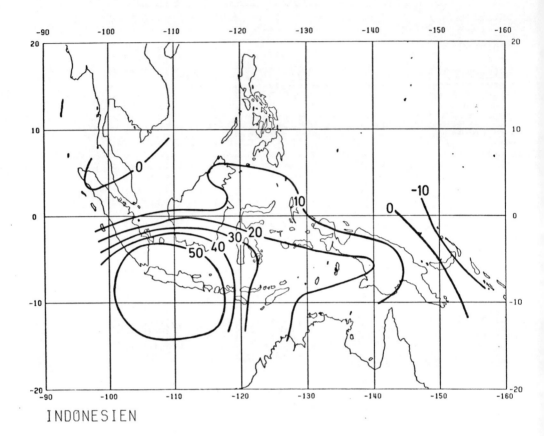

Fig. 3
Simultaneous isocorrelates (in %) of the RA of listed Indonesian rainfall series with basic series Java.

Moresby, Samarai, Kikori, Kokada, Kerema and Daru.

The same procedure repeated with the basic series Java led to an area-average for South Indonesia, now including thirteen stations marked in Table 2 with (S). Fig. 3 shows isolines of the average ccf with the basic series Java.

Finally the monthly amounts of rainfall of all stations lying in the corresponding region were averaged. If there was a gap in a particular month, the average for this month was taken.

b) Time series of other meteorological elements

In addition to the rainfall series the following time series (averages of individual months) were used:

1) 30 hPa zonal wind component near the equator from 1951 until 1977. Its source was Fig. 1 from Coy (1979), a time-height-plot of the stratospheric zonal wind component at Panama ($8°N$) until June 1970 and at Kwajalein from that time, after removal of the annual average.

2) SST Christmas Island ($1°59'N/157°22'W$). This time series was kindly supplied by K.-H. Weber, based on Seckel and Young (1971, 1977) for the time 1954-1974. The years 1975-1978 were obtained from SST series of Fanning Island ($3°55'N/159°32'W$) using linear regression (Vitousek, 1979).

3) SST North Pacific from 1947 until 1979. This time series is an area average over a large part of the North Pacific, ranging from $15°N$ to $60°N$ (about 45 million km^2) (Namias and Cayan (1981), Fig. 1).

4) SOI from 1852 until 1973, as given by Wright (1975), who published quarterly values of the SOI based on empirical orthogonal functions of six surface pressure stations, normalized with an average 0 and a standard deviation $\sigma = 1$. Wright computed SOI's for the quarters February-April (FMA), May-July (MJJ), August-October (ASO), November-January (NDJ); interpolated values have also been given for the meteorological seasons DJF, MAM, JJA and SON. The remaining quarterly values (JFM, AMJ, JAS and OND) were also interpolated linearly. This procedure yields approximate monthly values simply averaged over three consecutive months. The SOI can be interpreted as indicating the strength of the southeast trades over the Pacific.

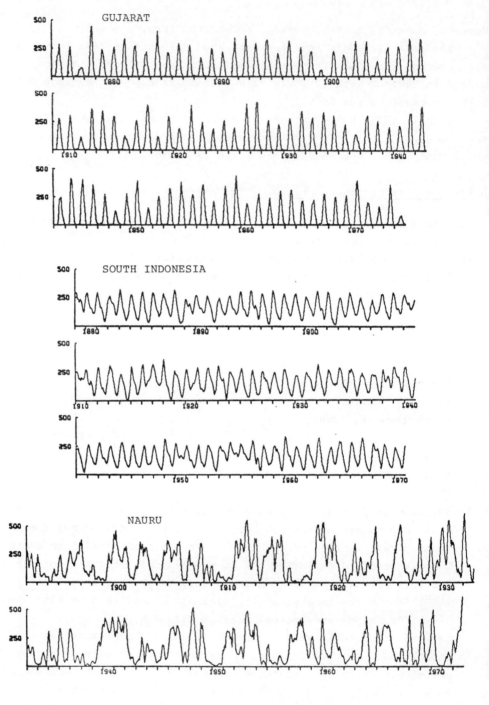

Fig. 4
Time series of selected rainfall stations and area averages. 3-months running-means taken for smoothing the time series.

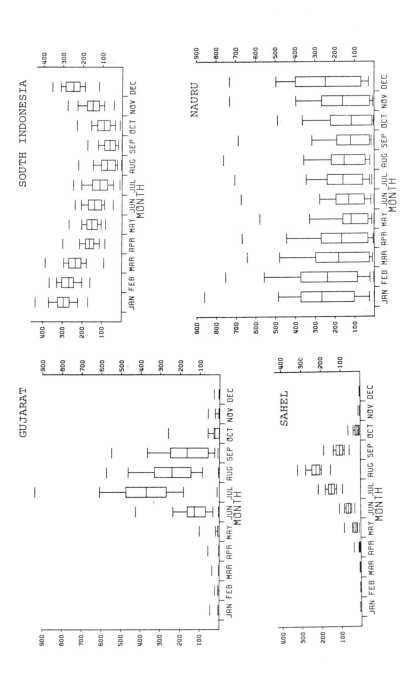

Fig. 5
Statistical data (maximum, upper decile, upper quartile, monthly average, lower quartile, lower decile, minimum) for selected rainfall series in dependence of calendar month. Vertical axis: rainfall in mm.

c) Properties of selected time series

In this subsection time series (Fig. 4) and some statistical data (Fig. 5) for selected rainfall series are shown; only a few remarkable properties shall be mentioned here. Interannual variations are rather distinct in South Indonesia, especially during the dry season of the SE-monsoon; we shall revert to this most important season later. Gujarat with its long dry season shows only variations during the southwest monsoon (June-September, Fig. 5). Most pronounced are the interannual variations at Nauru (Fig. 4) with probably the greatest variability of rainfall in the world. At this station the seasonal variations are much smaller than those correlated with the SO (Fleer, 1981). A more complete reconstruction of the Nauru series is given here by H.-P. Junk; the differences to the series used in this paper are negligibly small.

A.3 AUTOCORRELATION ANALYSIS

In this section an autocorrelation analysis is performed to identify seasons where RA's have a tendency to persist, also because of its role for estimating the significance of ccf's. In all series the serial ccf's of the RA were calculated for each month with time-lags up to 15 months. The results of a few selected series are shown in Fig. 6. Note that here running 3-months -RA's were taken for smoothing the isolines. Hence, the large acf's for lags up to two months are unrealistic. Other ccf's > 0.2 exceed the 90 percent confidence limit (Student's t-test).

In the Sahel and Gujarat series no persistence of their RA can be seen. In South Indonesia the persistence is strong between June and November, but weak or non-existent during the wet NW-monsoon (December-March). Nicholls (1981) explained this with a positive feedback mechanism between the anomalies of wind and SST during the second half of the year, which is negative during its first half. Much stronger is the persistence of the RA at the islands of the equatorial central Pacific, where it lasts for nearly all the year, i.e. from June to the following April (Fig. 6). Note the abrupt ending of the persistence in the northern spring (Wright, 1979). The SOI (Wright, 1975) and other rainfall series of the equatorial central Pacific show the same features. The reason for this phenomenon is as yet unknown, but it seems to be linked intimately with the onset and ending of El Niños in this region, which usually takes place in the northern spring (see section 5).

An interesting feature is the much stronger persistence of tropical RA compared with the RA in mid-latitudes. Fig. 7 shows this for selected rainfall series. Here the acf's are calculated again with monthly RA, but neglecting their seasonal dependence.

Fig. 8 shows the results of the autocorrelation analysis for each month (1-12). Here a "persistence time" was defined during which nearly all acf's within this range exceed the 90 percent confidence limit. The region East coast of Africa, for instance, has a persistence time from September to December. Here the acf's from September to October, November and December, from October to November and December, from November to December all exceed this confidence limit indicating a tendency for persistence between September and December, i.e. during the second rainy season. In Fig. 8 only those rainfall series are shown the persistence time of which lasts for more than two months. Like South Indonesia, Southeast New Guinea also shows a persistence time during the second half of the year. Central India shows a stronger persistence du-

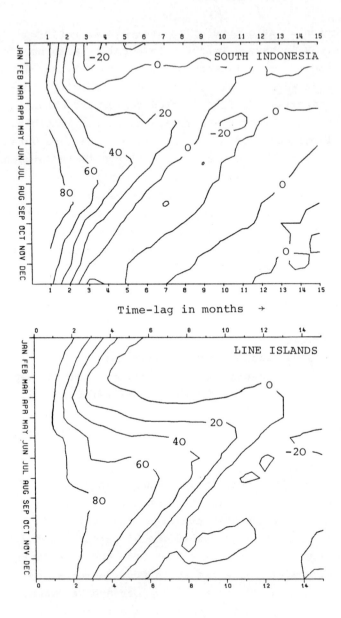

Fig. 6

Isocorrelates (in %) of quarterly RA centered at calendar month shown on ordinate, for two selected rainfall series.

ring the later part of the SW-monsoon, San Cristobal during the rainy season and Curaçao around the rainy season.

An interesting feature, common to most rainfall series, is the

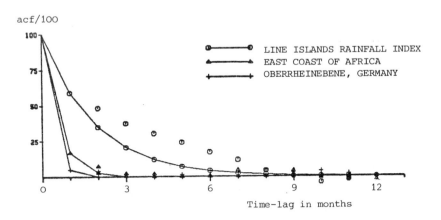

Fig. 7
Autocorrelation coefficients of monthly RA independent of calendar months in dependence of time-lag. Connected symbols denote autocorrelation coefficients from data evaluation, unconnected denote those of a Markov chain constructed with the rule: $ccf(n) = ccf(1)^n$. For comparison, curves for African and European climates are drawn.

great persistence around September. At this time many teleconnections are also stronger (cf. section 4).

Fig. 8
Length of time of persistence (months, ordinate) in dependence of calendar month (abscissa). Black bar for each calendar month shows, until which time-lag ccf of monthly RA exceeds 90% confidence limit, lower horizontal black bar denotes time of persistence as defined in text.

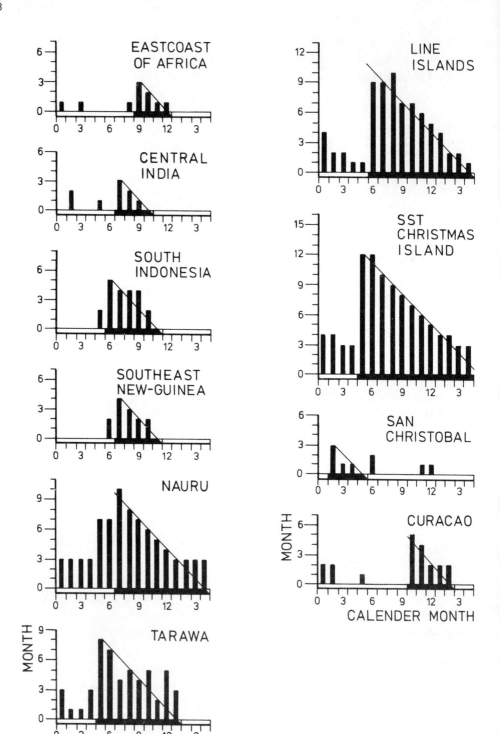

Fig. 8 (legend see preceding page)

A.4   CROSS-CORRELATION ANALYSIS

a) Calculations

For the identification of teleconnections two techniques have been widely used: analysis of cross-correlation and principal component analysis. Here the first method is preferred, since the time series used are of different length and time-lags up to $\pm$ 12 months had to be taken into account. Only such pairs of time series were used which have at least 20 years in common. To check the seasonal variability of the teleconnections, the ccf's were calculated in the following manner: each of the 12 subseries extracted from the basic series and consisting of the January data, the February data etc. was correlated with the subseries of the secondary series. With a time-lag of $\pm$ 12 months this yields 12x25 ccf's. They were arranged in matrix form (Fig. 9). These matrices were computed with the original data and also for running 3-months -RA's.

Due to the persistence of many series and to the great number of ccf's available for each pair, a frequency significance test was applied to all matrices with original data. If a ccf exceeds just the 90 percent confidence limit, its probability in a random series would be 10 percent. In other words: in every correlation matrix 10 percent of the ccf's may exceed the 90 percent limit without indicating a significant (linear) relationship. Thus significance can only be assumed if the number N of ccf's above the 90 percent limit is larger (Doberitz et al., 1967). Similar numbers are valid for higher confidence limits.

In highly persistent series, one method to take persistence into account is based on the equivalent persistence number $e(N)$ (Bartels, 1935) which depends on the autocorrelations (serial ccf's). Then an effective number $N_{eff} = N/e(N)$ can be estimated; this procedure raises the confidence levels and reduces the number of significant ccf's distinctly. In each selected area N was compared with N_{eff} for these confidence limits (90%, 95%, 99,5%). Within a given matrix a highly significant area (with a sufficient number of significant ccf's) was defined only if N exceeded N_{eff} at one (or more) of these limits.

The most interesting results of Fig. 9 are: significant positive ccf's from October (SON) until December (NDJ) at Sri Lanka with the highly persistent series of Nauru. These are significant at the 99.5 percent level. In marked contrast there is a series of negative ccf's

(significant only at the 90 or 95 percent level) between June (MJJ) and
September (ASO) with a time-lag to Nauru of about -5 to -12 months.
This apparent contradiction seems to be related to the varying coherence of the 5-year-wave between Nauru, Sri Lanka and parts of SE-India

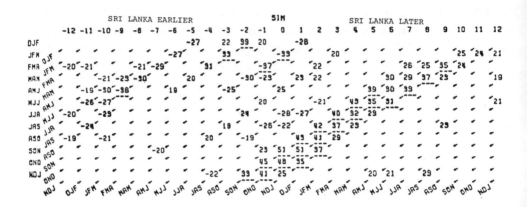

Fig. 9

Example of complete cross-correlation matrix, here between Nauru
(basic series) and Sri Lanka (secondary series). Numbers in uppermost line denote time-lag in months; the other numbers denote
ccf's exceeding 90% confidence limit in %. ccf's exceeding 95%
(99.5%) confidence limit plotted fat (underlined). Letters on
left and below matrix denote 3-months-periods (horizontal for basic series and diagonal for secondary series); e.g., DJF = December - February.

(Fleer, 1981, Fig. 20b) where several stations are in phase, others out
of phase with Nauru. Fleer's map and these results suggest that the
seasonal variations of rainfall - as caused by prevailing winds and
topography - result in a local modulation of the SO, which merits further investigations.

b) Results

The results are listed in the Appendix (p. 47); the main features only are described here. In the Appendix the regional teleconnections within large-scale regions with similar climates - such as the Indian subcontinent, the Indonesian Archipelago including SW New Guinea, the equatorial Pacific and northern Africa - are omitted; some results of these regional teleconnections are given below under i)-iv).

Also omitted are teleconnections occurring during an almost dry season, since these ccf's are crucially biased by a few large RA's. Furthermore, the rainfall amounts of such months are strongly non-Gaussian which does not allow to use them for estimating linear ccf's.

i) The RA's of Central India, Southeast India, Gujarat and North Pakistan show positive simultaneous teleconnections between each other during the SW-monsoon (June-September). They are strongest everywhere in September. During the second intermonsoonal season (October-November) the RA's of Sri Lanka and Southeast India are positively correlated; details have been given by Raatz (1977). Surprisingly, the Bangladesh series is not significantly correlated to the other series.

ii) The area-averaged RA's of the Indonesian Archipelago are strongly mutually correlated only in the second half of the year. South Indonesia shows the strongest large-scale teleconnections.

iii) The monthly RA's of Nauru, Tarawa and the Line Islands are strongly mutually correlated, also with the SOI and the SST-anomalies of Christmas Island (one of the Line Islands), for the greatest part of the year, i.e., from June to the following April. Simultaneous ccf's are mostly larger than 0.5; near the end of the year they even exceed 0.75. Only at the Line Islands a time-lag exists: here the RA's tend to follow those from Nauru by two months between November and April.

iv) The RA's of the Sahel and Sudan are mutually correlated only during the rainy season; no relation has been found to the SOI or other parameters used here.

The figures of this subsection concentrate upon the most important of these series, i.e., Central India, Sri Lanka and South Indonesia.

As expected, the correlations of the Indian RA's to other areas are strongest during the SW-monsoon, especially during September.

22

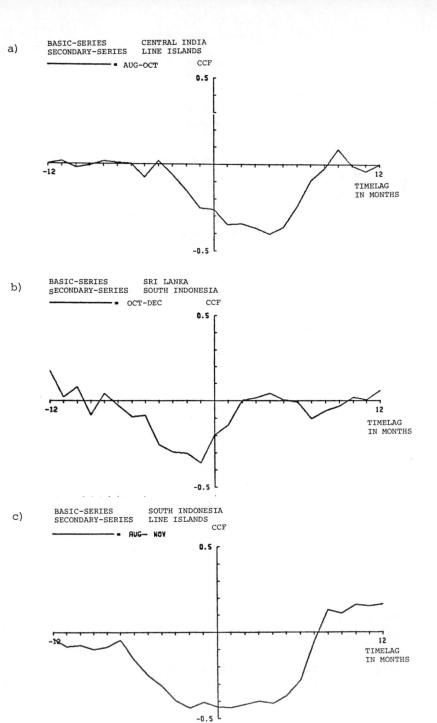

Fig. 10

Time-lags of teleconnections. Plotted are ccf's between selected periods (indicated in upper left) of basic series with secondary series.

d)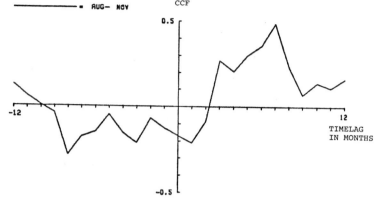

e)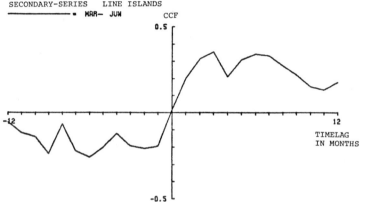

f)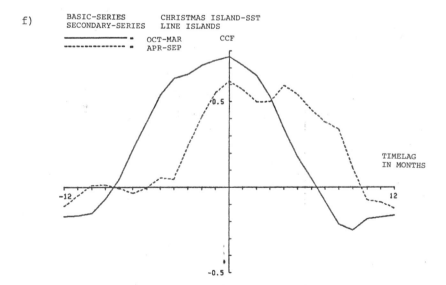

Monthly rainfall teleconnections towards Indonesia are weak (not more than about 0.4) but positive during the second half of the year, while those between India and the central Pacific Islands are negative and reach 0.5 (Mooley et al.) but last there from June to February because of the high persistence. The Line Islands -RA's lag about 3 months after India (Fig. 10a). Similar teleconnections are extended to the Galapagos (San Cristobal), here between the May-June-RA at India and the Galapagos rainy season 4 months later. Weak positive correlations even exist between India and the rainy season (autumn) at Curaçao, at a distance of $220°$ longitude, and also the Hawaiian winter rains 6 months later. In all these relations the Indian monsoon rains are leading, not lagging (Angell, 1981), as suggested 60 years ago by G. Walker.

The best teleconnections from Sri Lanka are found during the wettest season (October-November). They point during the same season to the same areas as those of the Indian SW-monsoon, but with opposite sign. Teleconnections to Indonesia are strong, with Indonesia leading (Fig. 10b). A weak positive correlation exists with the little rainy season (September-December) at the eastern coast of Africa.

In Indonesia the strongest teleconnections of the RA's are detected during the second half of the year, i.e., the season of strongest persistence. In contrast to Sri Lanka and Central India it is the dry season showing the highest correlations to the RA's of the equatorial central Pacific with no time-lag (Fig. 10c). Similar positive correlations are found to the Hawaiian winter rains and around the rainy season at Curaçao, both with a time-lag of 4 months. All these teleconnections are stronger than the same relationships from India.

From South Indonesia to the Galapagos, right across the Pacific (about $160°$ longitude), rather high positive correlations with a large time-lag (Fig. 10d) exist, but based on a short period only. Across the Indian Ocean weak simultaneous negative correlations are observed to the eastern coast of Africa. A rather unexpected result is the high correlation from the Marshall Islands (Kwajalein, leading by 4 months) to S. Indonesia, but also based on a short period.

Teleconnections from Kwajalein were taken into account only because of the importance of the NECC leading to an El Niño, as pointed out by Barnett (1977); no other series closer to the center of the NECC was available. It was assumed that the positive correlation between RA and SST detected in the upwelling region of the Pacific (Doberitz, 1968; Bjerknes, 1969) also holds here, and that this correlation may also represent the (varying) transport of warm water within the NECC.

Therefore the correlation matrices with Kwajalein as basic series were expanded to time-lags up to +24 months. Teleconnections from the RA's of Kwajalein were only found during northern spring where the NECC usually stagnates (Wyrtki, 1973). They point to the Galapagos a whole year later and to the Line Islands and Tarawa even 17-18 months later. These positive correlations reach values around 0.5. The lags are in good agreement with the time the water needs to flow from Kwajalein to Central America and to the South American coast, from where the SEC transports it again towards the west (Barnett, 1977). Direct (undelayed) correlations to the RA's of the equatorial central Pacific are of the same order, but negative.

The RA-teleconnections from the islands of the equatorial Pacific are strong in every direction. In addition to the above-mentioned results there are strong positive ccf's up to about 0.7 to the Galapagos towards the end of their rainy season and in the dry season (September-December), covering the second half of the year or even longer in the central Pacific. The RA's of these islands seem to lag after San Cristobal by a few months, but this lag is only significant within the correlations to the Line Islands. Additionally, there are strong negative teleconnections to the Hawaiian winter rains and (around the rainy season) at Curaçao. Because of the long persistence in the equatorial central Pacific time-lags within these teleconnections are not significant.

The correlations from the SST-anomalies of Christmas Island show the same features as those from the RA of the Line Islands. As expected, these two anomalies are very strongly correlated mutually and simultaneously (Fig. 10f). As already found by Doberitz (1968) and Ramage (1977), no time-lag between SST and RA can be detected within this area, at least not with the statistical methods used here.

No significant teleconnections could be found between our network and the area-averaged RA-series from North Africa, with the stratospheric (30 hPa) zonal winds and with the North Pacific SST. The latter series covers a giant area (45×10^6 km^2) which obviously is not coherent in itself, since not all of the spatial correlations within that area are significant (Namias and Cayan, 1981).

Last but not least we should fit these teleconnections into the SO by demonstrating correlation matrices with the SOI. Many correlations are detected, within the SOI-series, during the whole year, i.e., from May until the following April. As above, their strongest persistence is connected with only a few larger time-lags. In the following regions

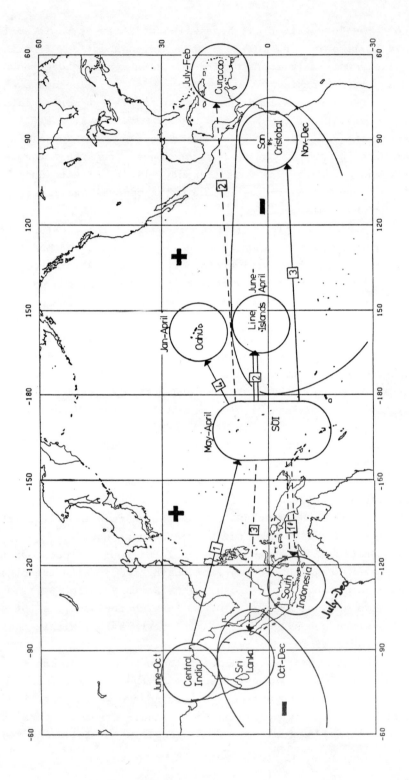

Fig. 11a
Teleconnections pointing to the SOI. Circles denote rainfall series and time ranges are the time ranges of the highly significant area of these rainfall series with the SOI. Thick sign denotes sign of ccf to SOI in time range shown to the surrounded rainfall series. Connecting lines between rainfall series and SOI indicate strength of maximum ccf and average time-lag of these teleconnections. The arrows on these connections show direction of time-lags.

——————— : $0.8 \leq$ max. ccf

---------- : $0.7 \leq$ max. ccf < 0.8

————— : $0.6 \leq$ max. ccf < 0.7

- - - - - - : max. ccf < 0.6

——[3]—— : time-lag = 3 months

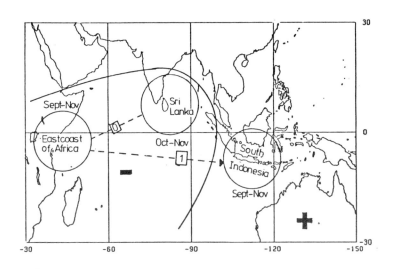

Fig. 11b)
Like Fig. 11a) but pointing to East coast of Africa.

the RA's show strong positive teleconnections to the SOI: Central India during the SW-monsoon, Kwajalein during the northern spring, San Cristobal only during the following dry season, Hawaii during the winter rains and Curaçao around the rainy season. Negative teleconnections exist to the RA's of the second intermonsoonal season at Sri Lanka, to the islands of the equatorial central Pacific from June until the following April, and to San Cristobal at the beginning of the rainy season. The teleconnections to Indonesia and the islands of the equatorial central Pacific are strongest; this had been expected since both areas are the centers of action of the SO (Berlage, 1957). But with one exception all other correlations reach or surpass 0.5. It should be noted that the simultaneously negative correlations to San Cristobal are weaker than the positive correlations with one year-delay, because rainfall in the Galapagos area is largely controlled by the El Niño phenomenon.

Fig. 11a summarizes the teleconnections with the SOI, Fig. 11b those in the Indian Ocean area. Fig. 12 takes the SOI during an "El Niño-epoch" (see section 5), computed for Fig. 15 and lasting 3 years, and shows the average RA of different rainfall series corresponding to this averaged SOI, as computed with the aid of linear regression. These rainfall anomalies are shown only for those seasons where the SOI reaches a highly significant area with the corresponding rainfall series and with the time-lag obtained from the cross-correlation analysis.

c) Discussion

Limiting our attention to the rainfall fluctuations (and their role in the tropospheric heat budget via release of latent heat), the main counterparts of the Walker circulation seem to be the maritime continent of Indonesia (Ramage, 1968) and the Line Islands of the equatorial Pacific from Nauru up to the Galapagos. This conjecture is based on the high ccf's of opposite sign in both areas between the RA's and the SOI. In some contrast to an earlier assumption (J. Bjerknes, 1969) our data suggest that the descending branch of the Walker circulation is best represented not by the Galapagos, but in the area of the Line Islands (see, however, the Appendix).

A second phase shift of the SOI between the equatorial Pacific and the long series of Curaçao (which is positively correlated within the Caribbean and some stations of northern South America, cf. Hastenrath and Heller, 1977, and Paper B (this issue)) is confirmed by the negative correlation between the RA's at the western coast of Peru-Ecuador and NE-Brazil (Coviedes). Thus the Pacific branch of the Walker circulation and its reversal on both flanks of the South American continent are verified by our results.

With a distance of 10 000 km between Indonesia and the east coast of Africa, the Walker circulation above the Indian Ocean is of smaller spatial scale. Its variations are documented in the weak negative RA-correlations between Indonesia in its ascending branch and the eastern coast of Africa and Sri Lanka in its descending branch. There are also high positive, simultaneous teleconnections between Sri Lanka and the equatorial Pacific islands, indicating a close connection between both Walker circulations, possibly because Indonesia is the ascending branch of both.

Since long rainfall series are only available from a few isolated areas, it cannot be expected that they represent the extreme positions of these circulations (Fig. 2). In addition to the limited spatial extension the season of the Walker circulation above the Indian Ocean seems to be limited: in Indonesia to the second half of the year, in Sri Lanka mainly to October-December, at the eastern coast of Africa to September-December. While the simultaneous correlations between SE-India and Sri Lanka are high (Raatz, 1977), the distant teleconnections with SE-India are remarkably weak. However, an analysis of the apparent 60-months-periodicity (Fleer, 1981) and its coherence with Djakarta and Nauru showed in both areas strange phase shifts at short distances. This result seems to indicate that, in spite of the significant spatial correlation within the selected areas, the use of not quite homogeneous area-averaged data can be misleading.

Looking at a reliable map of world rainfall, the area Indonesia-West Pacific receives more rainfall on a larger area than elsewhere, perhaps with one exception (western Amazonia). The climatic features of Indonesia are quite unique: a large area ($\sim 8 \times 10^6$ km^2) with large mountainous islands, producing, together with their diurnal circulations, enormous cumulonimbus systems shifting from land to sea (with maximum development around 16 h and 06 h local time, with unusually high SST (29-30°C) due to the absorption of solar radiation in the nearly cloud-free daytime hours). Taking into account the seasonal meridional shift of maximum rainfall in Indonesia, the annual maximum rainfall-area may occur above the western equatorial Pacific north of New Guinea; an example for the FGGE-year 1978/79 (Wei et al., 1983) shows this world maximum around 0°, 160°E with a diabatic heating of the atmosphere up to

30

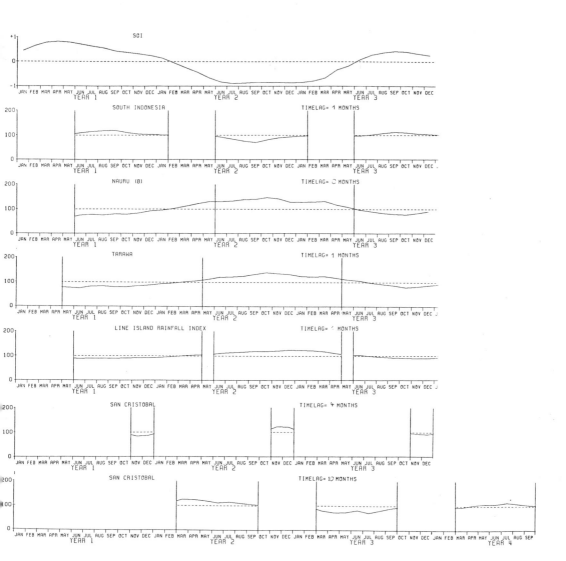

Fig. 12

Hypothesized RA for different rainfall series at given SOI and fixed time-lag. Positive time-lag means that for this rainfall series the RA occurs after corresponding anomaly of SOI. Ordinate shows RA in % of monthly average.
Dashed horizontal lines: Mean climatological rain.

120 W/m^2. As in our data the Indian Ocean branch of the Walker circulation is distinctly weaker than that of the Pacific, the occurrence of a second rainfall center during much of the year above NE-India contributes here to a more NE-SW directed thermal circulation.

Similarly, the teleconnections between the SOI and the RA's within the Pacific Walker circulation are much much stronger than within the Indian Ocean Walker circulation, in spite of the fact that the SOI had originally been designed for forecasting the Indian Monsoon. The following concluding picture of the teleconnections can be drawn: during the negative phase of the SO, when the Pacific southeast trades are weak, the Indian southwest monsoon and the relatively dry southeast monsoon in Indonesia are drier than normally, while unusually heavy rainfall occurs first at the Galapagos in the early rainy season of the year, reaching the equatorial central Pacific a few months later where they last a whole year. Similarly, the little rainy season at the eastern coast of Africa and the late-year rainy season in Sri Lanka receive more precipitation than usual. During the positive phase of the SO these RA's are of opposite sign.

Model computations, as quoted in section 1, are in good agreement with these results. They also confirm that the SO is in phase with the RA's at the equatorial Pacific islands, but not in the eastern Pacific (Keshavamurty, 1982). The spreading of a positive SST-anomaly from South America along the equator (Wells, 1979) is also indicated, providing that a positive RA can be taken as an indicator of a positive SST-anomaly. Keshavamurty (1982) also demonstrates the drought in India and Indonesia during the negative phase of the SO. However, no model has as yet demonstrated the teleconnections to Oahu, Curaçao, Sri Lanka and the eastern coast of Africa. The primary source of energy for driving the Walker circulation seems to be the release of latent heat (Corneja-Garido and Stone, 1977). In section 5 variations in the release of latent heat are presented for Nauru, Indonesia and India. There are strong interactions between the release of latent heat, the SST, the vergence of the wind, wind speed and direction. Large-scale variations in each of these meteorological elements can lead to variations of the intensity of the Walker circulation.

In the following the influence of a SST-anomaly on the teleconnections shall be considered. If a SST-anomaly is the forcing factor for a large-scale RA, large time-lags should occur between teleconnections of far distant regions. With an assumed current velocity of 0.5 m/s, a SST-anomaly needs about one year to cross the Pacific along the equator

(15 000 kms). In contrast to this, such time-lags cannot be assumed if atmospheric factors, primarily the release of latent heat, control these teleconnections; at a wind speed of about 8 m/s an air parcel needs only three weeks to cross the Pacific.

There are only small time-lags within the teleconnections between India and Indonesia on one side and the western Pacific islands Nauru and Tarawa on the other side. Together with the small SST-fluctuations in the Pacific west of the date-line (Rasmussen-Carpenter, their Fig. 22), this seems to indicate that SST-advection and -upwelling are not the primary factor in controlling the RA in this region. Nearly all teleconnections show the tendency to lag 1-3 months with respect to the Indian southwest monsoon; from here a great input of energy may reach the equatorial Pacific within weeks (see Angell, 1981). Similar small time-lags are also found within the Indian Ocean Walker circulation; this once more suggests that oceanic transport processes are not dominant. At larger distances significant time-lags are observed in the RA-correlations from the India-Indonesia center to the Line Islands and Galapagos and to Hawaii, Kwajalein and Curaçao. In the central and eastern Pacific the variations of the SST between El Niño-years and years of strong upwelling amount to more than 2 K (Ramage, 1975; see also Fig. 17). Such variations cannot be recognized in the equatorial Pacific west of the date-line and in the tropical Indian Ocean. This suggests that only east of the date-line SST-anomalies could be immediately responsible for large RA's in the equatorial Pacific.

The extremely strong rainfall variations at Nauru and other islands west of the date-line have been enigmatic for a long time; more detailed investigations are given in Papers C and D (this issue).

The frequency of westerly winds - see also the wind anomalies in the area given by Rasmussen and Carpenter (their Fig. 22) - suggests, during wet periods, a frequent occurrence of cyclonic eddies (or convergence zones) at both flanks of the equator, near $10°N$ and $10°S$ latitude, with westerly winds near the equator. The region showing similar RA's as Nauru apparently spreads farther west to the north of New Guinea; this is also indicated by low salinity (Donguy and Henin, 1978). In some contrast to this, the area where the sea-level is high during El Niño just reaches Nauru (Wyrtki, 1979) from the east. The high western Pacific-RA's during an El Niño could be induced by a positive SST-anomaly in the eastern and central Pacific, enhancing the Hadley cell to the west of it (Julian and Chervin, 1978) and hence the convergence of the trades. They are produced by a moisture convergence (Corneja-Garido and Stone, 1977) in this region, since in these latitudes both the ITCZ and the SPCZ approach each other near the equator in El Niño-years (Rasmussen and Carpenter, 1982), while the Indonesian low shifts to the date-line (Krüger and Winston, 1974; 1975).

In this context attention must also be drawn to the island of Kwajalein which is influenced by the NECC. The teleconnections from and to Kwajalein show large time-lags to the rainfall series in the equatorial Pacific. Important variations of the SST are not known, but Wyrtki (1973) stresses the variations in the transport of warm water and its influence on El Niños. Thus it can be assumed that the RA's are positively correlated with the intensity of the warm water transport. There are positive correlations to the Galapagos one year later and to Tarawa and the Line Islands another 5 months later. This coincides with the time necessary for advection of SST-anomalies from Kwajalein with the NECC to the Gulf of Panama, to the South American coast and, along the equator, back to the west, being in good agreement with Wyrtki's results. But these correlations are mainly limited to spring when the NECC is weak. In addition to these, nearly simultaneous negative correlations are found between Kwajalein and two of the equatorial Pacific islands, of similar magnitude. They suggest that atmospheric processes also contribute to the RA's of Kwajalein: the ascending branch of the Walker circulation is displaced along the ITCZ north of the equator, up to $170°E$ or even further.

In this discussion the variable release of latent heat by rainfall is considered as one of the most important sources of energy leading to the variations of the large-scale circulations such as the Walker circulation. If this is the primary factor in Indonesia and India, where the ascending branches of these circulations are usually located, it should be expected that the strongest teleconnections are caused in the wettest season.

Indeed, in India the SW-monsoon rainfall is significantly correlated with rainfall in the equatorial West and Central Pacific. However, no significant correlation could be found with the center of maximum rainfall in Bangladesh (where the interannual variations are only small). In West Indonesia very few correlations are found with Pacific rainfall during the first half of the year, especially during the NE-monsoon (January-April). In contrast to this, significant and high correlations exist between the relatively dry season of the SE/SW-monsoon, especially around September - the driest month. A similar seasonal preference is found in New Guinea, in Sri Lanka (where the late-year intermonsoon-period dominates) and even along the African east coast. The opposite results from Kwajalein and the Galapagos (here rainfall is limited to January-April) are much less representative.

Thus the strongest teleconnections are generally found, in the key areas, during the second half of the year, especially in September/October. This can tentatively be interpreted as a consequence of the intensity and steadiness (Troup, 1965) of the SE-trades extending farthest across the equator into the northern hemisphere.

The intensity of the SE-trades is positively correlated with the SOI (Quinn and Burt). It should be remembered that the persistence of many of our RA-series is highest in September (section 3); only in the central Pacific it is higher during the first half of the year but with a secondary maximum during September. From an oceanographic point of view Wyrtki (1975, 1977) also stressed the role of the Pacific SE-trades for the origin of El Niño-situations.

The remaining results can only briefly be mentioned:

1) The relatively weak teleconnections towards the Hawaiian winter rains are treated by Meisner (1976).
2) Curaçao is situated in a regional upwelling-area where the speed of the ESE-trades increases towards W (Becker, 1982).
3) No significant teleconnections between the RA's of the Sahel-Sudan-belt with parameters of the SO could be found, perhaps due to its continental position.
4) No correlations are found between the RA's of the equatorial Pacific and the stratospheric biannual oscillation, represented here by the zonal wind component at 30 hPa. This coincides with the results of power spectrum analyses of these RA's (Fleer, 1975): the contribution of the biannual period to their total variance is much smaller than that of a 5-year-quasiperiod.
5) Shukla (1975) proposed with his model a negative correlation between Indian summer monsoon-RA and that of the East African coast; this could not be verified by our long and homogeneous data series. It should be noted that the spatial distribution of rainfall on the Indian subcontinent is not uniform; our series are limited to a coherent net of stations (Schweitzer, 1978).

A.5 COMPOSITE ANALYSIS OF EL NIÑO RAINFALL ANOMALIES

With a composite technique as used by Rasmussen and Carpenter (1982), the RA's of selected rainfall series and the SOI are treated before, during and after an El Niño. Then it is possible to put the El Niño-anomalies into the scheme of the teleconnections found in section 4 and to investigate the displacements and the magnitude of the RA's during the variations of the Pacific Walker circulation (or the SOI) in different regions. For this purpose "El Niño-Epochs" lasting 3 years from January before until December after an El Niño-year are taken. Since 1895 the following years have been selected as El Niño-years: 1899, 1905, 1911, 1918, 1925, 1930, 1941, 1951, 1953, 1957, 1965, 1969, 1972 and 1976 (Wooster and Guillan, 1974; Fleer, 1981). A number of separate El Niño-epochs was used to compute, for each rainfall series, a composite run of 36 monthly average-RA's, expressed in percent of the long-term average. Here it is assumed that the El Niño and other meteorological events related to it occur in the same season (Rasmussen and Carpenter, 1982). In each composite the monthly values are smoothed by a 3-monthly running mean except for the marginal values. The rainfall series used are listed in Table 1 except for Fanning Island (see section 2), Washington Island ($4°43'N/160°25'W$, also one of the Line Islands) and Canton Island ($2°46'S/171°43'W$).

Figs. 13 - 15 and 18 show the composites with average SOI-values (left scale) and the rainfall series in percent of the average (right scale) versus the El Niño-epoch. In the right upper corner the El Niño-years used are listed. Along the composite RA-lines crosses are used if the normalized standard deviation σ is <0.5, while triangles (circles) indicate σ-values between 0.5 and 1.0 (>1.0).

Fig. 13 shows a zonal cross-section along the Pacific equator from the Galapagos ($89°W$) to Nauru ($167°E$) for the El Niños between 1950 and 1973. Due to incomplete data, the El Niño time-range 1971-1973 ends at Nauru and Christmas Island in 1972; this inhomogeneity is of but little importance in the last year. In the "Prae-El Niño-year" the SOI is positive and all RA's stay below 100 percent, except for a few months at San Cristobal. At Nauru and Tarawa less than 50 percent are observed in most of the months. During the El Niño-year the average RA reaches more than 100 percent at Nauru and San Cristobal in January, at Tarawa in February and at Fanning Island in April, simultaneously with the highest peak at San Cristobal. The SOI raches its minimum in July. After the short, rather constant rainy season at San Cristobal in the first

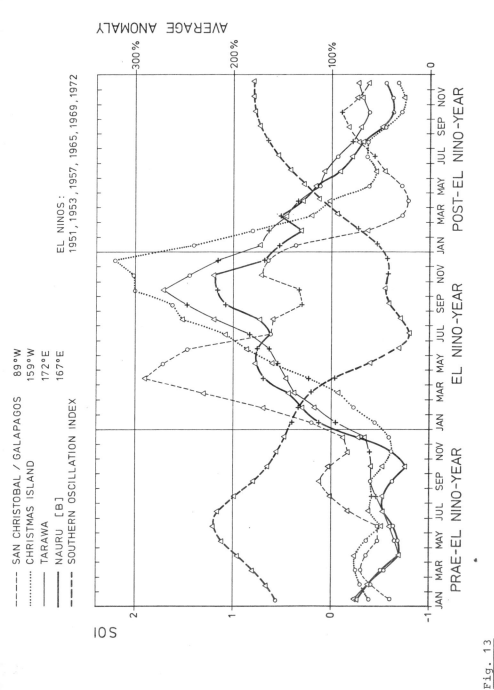

Fig. 13 Composites of RA's at stations San Cristobal, Christmas Island, Tarawa, Nauru, and the SOI.

half of the year the maximum rainfall at the central Pacific Islands occurs during the later months of the year (September-December); here the absolute rainfall amounts are about equal at all three stations. During the Post-El Niño-year the RA's are rapidly decreasing, proceeding from east to west.

Fig. 14 shows (for four El Niño epochs only, i.e., 1950-1966) in a meridional cross-section at $160°W$, including the Line Islands and Canton Island, a composite of the RA's and the SOI. This figure suggests that during an El Niño the heavy rainfalls spread meridionally on both sides of the equator. They sucessively reach Christmas Island (next to the equator), Canton Island (south of the equator), Fanning Island and at last Washington Island, at the greatest distance from the equator. The peak occurs at all stations nearly at the same time, while its relative intensity decreases from south to north. At the beginning, the disappearance of rainfall seems also to spread in a poleward direction. These features remain unchanged if for the three Line Islands (except Canton) composites for six El Niño-epochs (1950-1973) are given (not reproduced here).

Using not less than 12 El Niño-epochs between 1898 and 1970, composites of Central India, South Indonesia, Curaçao and the SOI are shown in Fig. 15. The composites of Nauru and of the SOI are rather similar to those shown in Fig. 12 for 6 cases only, but during the El Niño-year the extreme values are larger. At Curaçao the RA's are positive in the Prae-Niño-year, but negative during the El Niño-year, with values of about 50 percent from September until the following April. At the end of the Post-Niño-year they reach again a maximum.

In Central India, area-averaged rainfall exceeds the normal values by 30-40 percent from August to October in the Prae-Niño-year, but drops to about 70 percent of the normal values from July to October in the real El Niño-year, thus confirming the well-known seesaw between the West Pacific and India. In southern Indonesia rainfall is subnormal from November before until the end of the El Niño-year, i.e. during a period of 14 months, with deviation up to 40 percent in December and January.

Fig. 16, on a longitude-time plane along the equator ($8°S-12°N$), i.e., omitting Central India, demonstrates the large-scale RA-variations over a longitudinal distance of $260°$. A notable exception is San Cristobal (Galapagos) where only dates from 1950-1980 were available; dashed isolines are used here to delineate the inhomogeneity of the data source. In the Pacific the anomalies are stronger than in the Indian

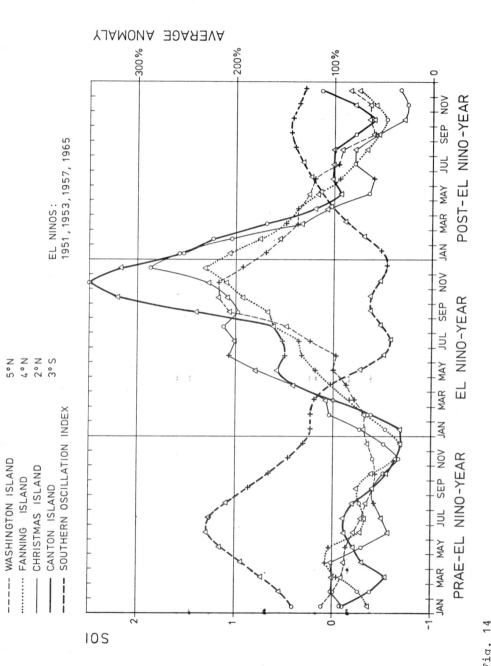

Fig. 14 Composites of RA's at stations Washington Island, Fanning Island, Christmas Island, Canton Island, and the SOI.

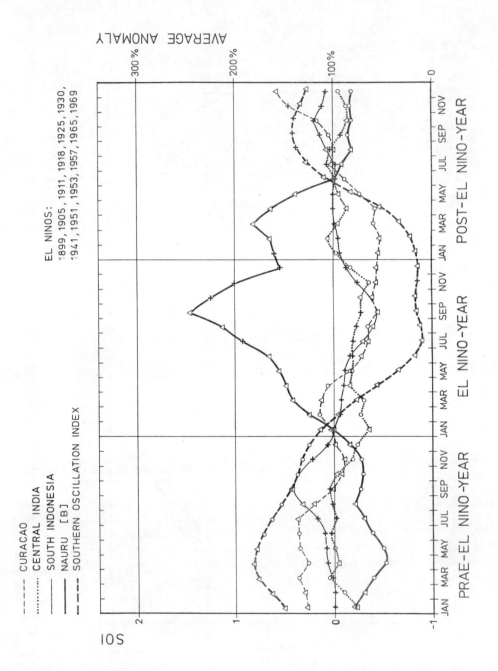

Fig. 15   Composites of RA's at stations Curaçao, South Indonesia, Central India, Nauru, and the SOI.

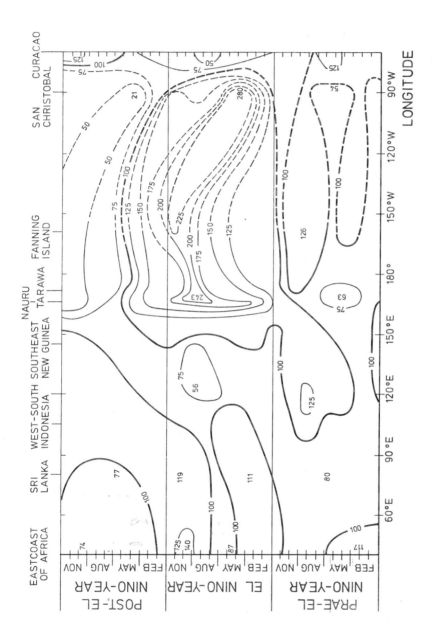

Fig. 16 Longitude time plot of RA (%) on equator during an average El Niño.
Dashed: Subjectively interpolated.

Ocean area and in the Caribbean; most remarkable during the El Niño-year is the phase shift near 150°E and the rapid increase of RA's towards the Pacific. Here the rains start early in the year (in northern spring) at the Galapagos as well as at Nauru, while the central Pacific (Line Islands) maximum is delayed until the end of the year. Until now, little attention has been given to the negative anomaly over the Pacific during the Post-Niño-year, apparently extending also over about 120° longitude.

The region with positive SST-anomalies during the El Niño-year (Rasmussen-Carpenter) coincides well (also in time) with the positive rainfall values; the seasonal delay in the central Pacific can be recognized in a weak form. Between 120° and 150°E the sign of the SST-anomaly becomes weakly negative; the same is true over the Pacific during the main part of the Post-Niño-year, both in accordance with the RA's. Most striking are the westerly wind anomalies between 150°E and the Line Islands (near 150°W). This is accompanied (not shown here) by an eastward displacement of the low pressure area from Indonesia (see the discussion in section 4).

In this context Table 3 should be mentioned: it shows the variations of latent heat release by rainfall in 5 months-periods during two adjacent years of an El Niño-epoch. The station record of Nauru is

Table 3
Variations in the release of latent heat between the listed seasons in El Niño-years and Prae-El Niño-years

Rainfall series	Nauru	South Indonesia	Central India
Season	July-Nov	July-Nov	June-Oct
Prae-El Niño-year	89 W/m^2	118 W/m^2	133 W/m^2
El Niño-year	273 W/m^2	57 W/m^2	102 W/m^2
Difference	184 W/m^2	61 W/m^2	31 W/m^2

certainly not representative for the whole equatorial wet region of the western Pacific which may enclose some 50° long. x 10° lat. or about 7×10^6 km^2; but South Indonesia represents also about 3.3×10^6 km^2.

Both data indicate that the heat released into the atmosphere over large source-regions of large-scale zonal (Walker) circulation can vary interannually by a factor of at least 2, simultaneously with a displacement of the heating centers over about 4000-6000 km.

Fig. 17 shows once more the possibly contributing role of the NECC to an El Niño, using 6 epochs between 1950 and 1973; again 1973 is missing at Christmas Island. The weak RA's at Kwajalein are positive du-

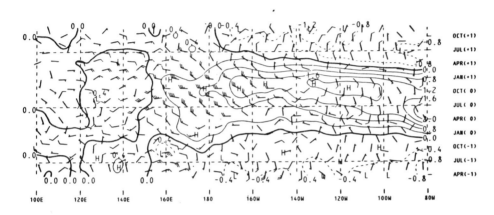

Fig. 17

Longitude-time plot of wind- and SST-anomalies during an average El Niño. (Adopted from Rasmussen and Carpenter, 1982.)

ring nearly a whole year before an El Niño. 11 months later this signal reaches the Galapagos with a start of an El Niño, and after a further delay of three months (8 months until the peak) the anomaly reaches Christmas Island, though this station is 1700 km nearer to Kwajalein than San Cristobal. However, due to the lack of further data in the NECC-region, this conjecture needs verification.

A.6 CONCLUDING COMMENTS

The results of our composite analysis of rainfall anomalies covering about two thirds of the earth's circumference along the equator agree quite well with those obtained by Rasmussen and Carpenter (1982), mainly based on SST's and low-level winds. As an example: during an El Niño-year - i.e., during the negative phase of the SOI - weak positive SST-anomalies reach Nauru (long. 167°E) a few months earlier than the Line Islands (long. 158°W), similar to the RA's. As already suggested by Doberitz (1968), both RA's and SST-anomalies apparently spread polewards from the equator with a speed near 1° lat. per month (about 4 cm/s). During an El Niño, this effect could possibly originate in the absence of upwelling cool water at the equator. Normally, the diverging Ekman-drift drives cool water upwards along the equator which spreads polewards; in the Atlantic (see Paper C of this issue) this suppresses convective activity also at lat. $0-2^{\circ}$N. With upwelling being replaced by downwelling and rising SST, convective rainfall tends also to spread polewards.

This effect could also be responsible for the 4 months-delay of the onset of RA's at the Line Islands (lat. $2-6^{\circ}$N) compared with the Galapagos (Fig. 18). Thus the fictitious spread of the RA's from east to west - which could also not have been recognized by Sadler (1969) using satellite data from the 1965 El Niño - could be governed by a poleward spread of $SST > T_{air}$ and thus by convection. This could verify Wyrtki's (1975, 1979) hypothesis proposing that the SST-anomalies change their sign more or less simultaneously along the equatorial Pacific when equatorial Kelvin waves excited at the western Pacific travel eastward to South America. In contrast to this initiation of rainfall the RA-maxima seem to be forced by the advection of warmer water, partly contributed also by the NECC (cf. the 11 months-time-lag between the positive RA's at Kwajalein and the Galapagos).

The composites show also the abrupt ending of the autocorrelations of the RA's within the Line Islands (see section 3): during El Niño the RA's rise above the average in northern spring, but drop below normal exactly one year later.

During the finalization of this paper (winter and spring 1982/83) a new dramatic El Niño-event occurred in the Pacific region in an unusual season and with unprecedented intensity, accompanied by many unexpected anomalies nearly all over the globe. It was agreed not to consider this 1982/83-event, due to the lack of a comparable data base;

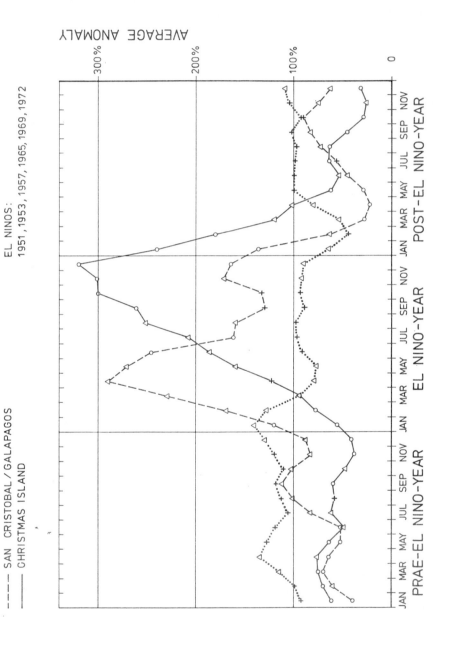

Fig. 18 Composites of RA's at stations Kwajalein, San Cristobal and Christmas Island.

meanwhile first comparative investigations have been published (M.A. Cane, 1983, E.M. Rasmussen and J.M. Wallace, 1983). Some experience with earlier cases (see Paper C of this issue) seems to indicate that in the western and central Pacific El Niño-events may occur practically at any time, while at the South American coast a seasonal preference is observed. Similarly the correlations between SOI and the Indian Summer Monsoon-RA's depend on season: here northern summer, in contrast to the El Niño rainy season at the Ecuador-Peru coast. These facts seem to indicate that as near-global phenomena the El Niño-events cannot be considered as seasonally fixed. Apparently such events in the western/central Pacific are more frequent than their counterparts in far distant coastal and continental regions. This and the extreme extension of the phenomenon need more attention: here we see initiation and consequences of climatic anomalies working under our eyes, on the largest possible scale.

APPENDIX: Significant cross-correlations

The following table lists the highly significant areas of the cross-correlation matrices of section 4; ccf denotes the largest ccf of the cross-correlation matrix the input of which are running 3-months-RA's, τ the time-lag of those ccf's; here a positive (negative) sign indicates that the basic series leads (lags) the secondary series, BS (SS) is the central month of the basic series (secondary series) with the highest ccf. Within the highly significant area (Jan-Feb) x (July-Aug), e.g., means that the time-range January-February of the basic series and the time-range July-August of the secondary series contribute to the highly significant area. n denotes the number of pairs.

Basic series Gujarat

Secondary series	ccf	τ	BS	SS	highly significant area	n
Oahu	+35	+4	Oct	Feb	(Sept-Oct) ✕ (Jan -Feb)	84
San Cristobal	-56	+3	Sept	Dec	(Aug -Oct) ✕ (Dec -Jan)	25
Sudan	-40	-11	Aug	Sept		72

Basic series Central India

	ccf	τ	BS	SS	highly significant area	n
South Indonesia	+38	+3	Sept	Dec	(Aug -Oct) ✕ (Aug -Jan)	92
Nauru	-48	+1	Aug	Sept	(June-Oct) ✕ (July-Feb)	81
Line Islands	-58	+3	Oct	Jan	(June-Oct) ✕ (July-Feb)	65
San Cristobal	-69	+4	Aug	Dec	(July-Sept) ✕ (Dec -Jan)	25
"	-54	-2	Aug	June	(Aug -Sept) ✕ (May -June)	25
Curacao	+36	+5	June	Nov	June ✕ (Sept-Nov)	76

Basic series Southeast India

	ccf	τ	BS	SS	highly significant area	n
South Indonesia	+43	+3	July	Oct	(July-Aug) ✕ (Aug -Dec)	92
SE New-Guinea	-40	-3	Dec	Sept	(Oct -Dec) ✕ (Aug -Sept)	69
Oahu	+41	+6	July	Jan	(July- Aug) ✕ (Jan Feb)	76
Curacao	+39	+7	Aug	Mar	(July-Aug) ✕ (Jan -Mar)	76
"	-38	+3	Nov	Feb	(Oct -Nov) ✕ (Feb -Mar)	76
Sudan	+42	-7	Feb	July	(Jan -Feb) ✕ (July-Aug)	72

Basic series Sri Lanka

	ccf	τ	BS	SS	highly significant area	n
West Indonesia	-37	-1	Oct	Sept	(Oct -Nov) ✕ Sept	92
"	+37	-5	Aug	Mar	(July-Aug) ✕ (Feb -Mar)	92
South Indonesia	-57	-2	Oct	Aug	(Oct -Dec) ✕ (June-Nov)	92
SE New-Guinea	-53	-1	Oct	Sept	(Oct-Nov) ✕ (July-Nov)	69
Nauru	+53	0	Oct	Oct	(Oct -Nov) ✕ (June-Dec)	81
Tarawa	+58	-2	Oct	Aug	(Oct -Nov) ✕ (June-Nov)	50
Line Islands	-42	+6	July	Jan	July ✕ (Dec -May)	66
"	+50	-1	Oct	Sept	(Oct -Nov) ✕ (June-Mar)	66
Oahu	-34	+3	Oct	Jan	Oct ✕ (Jan -Mar)	85
San Cristobal	-63	+6	Oct	April	(Oct -Nov) ✕ (Mar -May)	26
"	+64	0	Nov	Nov	(Nov -Dec) ✕ Nov	26
Curacao	-41	+3	Nov	Feb	Nov ✕ (Jan -Mar)	76
Sudan	+38	+5	Mar	Aug	(Feb -Mar) ✕ (July-Aug)	72
Eastc. of Africa	+42	+1	Oct	Nov	(Oct -Nov) ✕ (Sept-Dec)	79

Basic series West Indonesia

secondary series	ccf	°C	BS	SS	highly significant area	n
Nauru	-54	0	Aug	Aug	(July-Nov) ✕ (June-Dec)	79
"	+38	-4	May	Jan	(April-June) ✕ (Dec-Jan)	79

Basic series South Indonesia

Kwajalein	+67	-4	Sept	May	(Aug -Nov) ✕ (April-May)	26
Nauru	-66	+2	Sept	Nov	(June-Jan) ✕ (April-Jan)	79
Tarawa	-57	+1	Oct	Nov	(April-Nov) ✕ (May -Dec)	45
Line Islands	-66	-3	Oct	July	(Aug -Nov.) ✕ (May -Mar)	61
Oahu	+48	+4	Sept	Jan	(Aug -Jan) ✕ (Jan -Feb)	80
San Cristobal	+65	+4	Nov	Mar	(Aug -Nov) ✕ (Mar -May)	21
Curacao	+50	+4	July	Feb	(June-Sept) ✕ (Oct -Mar)	76
Eastc. of Africa	-41	-1	Nov	Oct	(Sept-Nov) ✕ (Aug -Nov)	78

Basic series Southeast New-Guinea

Nauru	-62	+1	Sept	Oct	(July-Nov) ✕ (June-Mar)	69
Tarawa	-59	-3	Sept	July	(July-Nov) ✕ (April-Feb)	47
Line Islands	-58	0	Nov	Nov	(July-Dec) ✕ (Aug- Mar)	63
Oahu	+44	0	Jan	Jan	(Aug -Jan) ✕ (Jan -Feb)	69
Curacao	+48	+2	Nov	Jan	(Nov -Dec) ✕ (Sept-Feb)	67

Basic series Kwajalein

Tarawa	-61	-1	Mar	Feb	(Feb -Mar) ✕ (Jan-April)	36
"	+47	+17	May	Oct	(Mar -June) ✕ (Sept-Nov)	36
Line Islands	-56	+1	Mar	April	(Feb -Mar) ✕ (April-May)	31
"	+59	+18	April	Oct	(April-May) ✕ (Aug -Oct)	31
San Cristobal	+72	+5	Feb	July	(Jan -Mar) ✕ (July-Aug)	31
"	+47	+11	May	April		31

Basic series Nauru

Oahu	-55	+3	Nov	Feb	(June-Feb) ✕ (Jan -Mar)	81
San Cristobal	+77	-3	Sept	June	(April-Dec) ✕ (April-June)	26
"	+59	+1	Oct	Nov	(Oct -Nov) ✕ (Oct -Nov)	26
Curacao	-57	0	Dec	Dec	(June-Feb) ✕ (July-Mar)	76

Basic series Tarawa

San Cristobal	+74	-3	Sept	June	(May -Jan) ✕ (April-June)	31
"	+68	-1	Dec	Nov	Dec ✕ (June-Dec)	31
Curacao	-66	+2	Nov	Jan	(Sept-Mar) ✕ (Sep-Feb)	45
SST Chr. Island	+83	-1	Feb	Jan	(July-Mar) ✕ (July-Mar)	25

Geographisches Institut
der Universität Kiel
Neue Universität

Basic series Line Islands

secondary series	ccf	τ	BS	SS	highly significant area	n
SST Chr. Island	+93	0	Nov	Nov	(June-April) × (June-Apr)	24
Oahu	-62	0	Mar	Mar	(July-April) × (Jan-April)	66
San Cristobal	+62	-4	Oct	June	(June-Nov) × (Mar -June)	26
"	+64	0	Nov	Nov	(July-Mar) × (Sept-Dec)	26
Curacao	-68	+1	Dec	Jan	(July-Mar) × (Aug -Feb)	61

Basic series SST Chr. Island

San Cristobal	+79	-3	Jan	Oct	(Aug -Mar) × (Sept-Dec)	25
"	-78	+9	Oct	July	(April-Dec) × (May -Aug)	25

Basic series Oahu

San Cristobal	-61	-8	Oct	Feb	(Sept-Oct) × (Feb -Mar)	26

Basic series SOI

North Pakistan	+50	-1	Oct	Sept	(May -Jan) × (Aug -Sept)	61
Gujarat	+44	-1	Oct	Sept	(July-Mar) × (July-Oct)	99
Central India	+60	-1	Oct	Sept	(May-April) × (June-Oct)	104
Sri Lanka	-57	+3	July	Oct	(April-Dec) × (Oct -Dec)	99
West Indonesia	+56	0	Aug	Aug	(May-April) × (July-Sept)	92
South Indonesia	+73	+1	Aug	Sept	(April-Mar) × (June-Jan)	92
SE New-Guinea	+59	+3	July	Oct	(May-April) × (Aug -Dec)	69
Kwajalein	+56	0	April	April	(Mar -May) × (Mar-April)	29
Nauru	-78	0	Nov	Nov	(May-April) × (June-May)	81
Tarawa	-72	+2	Nov	Jan	(May-April) × (May-April)	48
Line Islands	-85	+2	Nov	Jan	(May-April) × (June-April)	64
SST Chr. Island	-91	0	Oct	Oct	(May-April) × (June-April)	20
Oahu	+65	+4	Oct	Feb	(May-April) × (Jan-April)	83
San Cristobal	-63	+3	Aug	Nov	(July-Dec) × (Nov -Dec)	24
"	+73	+9	Oct	July	(April-Nov) × (Mar -Sept)	24
Curacao	+59	+2	Nov	Jan	(May -Mar) × (July-Feb)	76

B   Cornelis Becker und Hermann Flohn:

   TELECONNECTIONS IN THE
   CARIBBEAN AND NORTHERN
   SOUTH AMERICA AND THE
   SOUTHERN OSCILLATION

ABSTRACT

Simultaneous correlation coefficients between rainfall anomalies and the Southern Oscillation (as well as with rainfall in Nauru) are presented. The anomalies are on the half-yearly and seasonal time-scale. The rainfall data are from selected stations in northern South America. The correlation is in most cases highly significant. The regional coherence between the South American stations is demonstrated with simultaneous monthly correlation coefficients.

In his diploma thesis (University of Bonn, 1982) Cornelis Becker (Surinam) has investigated, on the basis of 15 available monthly and annual long rainfall series from the Caribbean and northern South America, teleconnections of RA's[*)] together with their relation to the SOI[*)] (Wright, 1975) and to Nauru rainfall as another parameter representing SO. Unfortunately, the distance of most stations was so large, that the formation of a coherent area-average proved to be of little value. The only area where such an area-average could have been justified was at the northern South American coast between long. $55°$ and $62°W$, with the stations St. Clair (Trinidad), Georgetown (Guyana), Nw. Nickeri ($5°51'N$, $57°2'W$) and Paramaribo (both Surinam); here simultaneous monthly RA's are correlated up to $r = 0.83$. However, long rainfall series from single stations frequently lack homogeneity. This prohibited the use of such important stations as Ancon (Ecuador, $0°13'S$, $78°30'W$) or Recife (Brazil, $8°06'S$, $34°54'W$); also Trinidad seems to be inhomogeneous before the year 1900.

The drought-stricken area of northeastern Brazil is represented only by the rainfall record from Quixeramobim ($5°12'N$, $39°18'W$, available 1896-1960, annual average 689 mm); the much longer record from Fortaleza at the coast (since 1849) was only available in annual averages (Markham, 1974). Earlier investigations (Eickermann and Flohn, 1962, Doberitz, 1969, Caviedes, 1973, Hastenrath and Heller, 1977) have found a negative correlation between SST[*)] and RA's at the coast of Peru/Ecuador (or in the eastern and central Pacific) and those in NE-Brazil, in northern South America (Berlage, 1957), in the Caribbean (Reiter, 1983, Fig. 8) and in the equatorial Atlantic.

Becker's investigation was based on the same techniques of correlation analysis as Behrend's (paper A of this issue). Here only a selected number of results shall be given, which indicate that the use of Curaçao as a representative station for northern South America is justified, in spite of the peculiar dryness of these islands. The important problem of NE-Brazil cannot be considered in detail, due to the unavailability of sufficient data; reference should be made to recent investigations (Hastenrath and Kaczmarczyk, 1981; Chu, 1983).

Table 1 gives a selection of ccf's between seasonal RA's[*)] in northern South America and the SOI (Wright, 1975), limited to simultaneous values. Similar ccf's with a time-lag of $\pm$ 1-4 seasons are given by

---

[*)]
 For acronyms see footnote on p. 3.

Becker; due to the high persistence of SOI a certain number of significant ccf's exists - at least for lags up to ± 2 seasons (6 months). The 99 year-series of Trinidad shows, for each season, weak but significant ccf's (0.22-0.32), which are apparently biased by the inhomogeneity of the record around 1900. All other ccf's in Table 1 are significant at

Table 1
Simultaneous ccf's (significant at the 90%-level or higher; unit 0.01) between SOI and 3-months-RA's (n = number of years)

	n	FMA	MJJ	ASO	NDJ
Bogotá	67	40	52	29	--
Curaçao	76	28	--	45	51
N. Nickeri	70	--	22	48	33
Paramaribo	75	--	29	32	31
Caracas	70	--	--	40	29
St.José (Costa Rica)	77	--	31	52	31
Quixeramobim	65	41	--	-26 (!)	32

the 90 percent level (or higher); as in other areas they reach highest values during the second half of the year. This is not the case, however, for Bogotá, at an altitude near 2600 m near the meteorological equator (lat. 5°N), where the highest ccf's occur during the first half year.

Table 2 presents a selection of simultaneous half-year correlations between South American stations and Nauru. Most of them are highly significant for the two adjacent half-years, and all show a phase shift between central Pacific and northern South America, i.e., over a zonal distance of 106-134° long. or more than 12 000 km..

Within northern South America the spatial coherence of these RA's can be remarkable. As examples of simultaneous half-year intervals the two high altitude-stations Bogotá and Quito are correlated with r = 0.45 (6-11) and 0.56 (12-5); for the same intervals the ccf's for Curaçao-Caracas are 0.56 and 0.65, for Curaçao-Trinidad 0.39 and 0.61, all significant at the 95 percent level. Table 3 gives a selection of monthly ccf's between Curaçao and other stations of the region. As in many other tropical areas (see Nicholson (1983, Fig. 6) for Africa, Nicholls (1981) and paper A of this issue for Indonesia) the spatial coherence of RA's is quite impressive and obviously higher than in mid-

Table 2

Simultaneous ccf's (unit 0.01) between Nauru's and other stations' RA's (6 months, 5-10 = May-October); significance level 99.5% (95%).

	5-10	6-11	7-12	8-1	9-2	11-4[1)]
Bogotá	-50	-44	-45	-47	-42	-46
Curaçao	-49	-44	-52	-58	-59	-63
N. Nickeri	(-28)	-37	-50	-42	-33	(-23)
Georgetown	-34	-44	-45	-53	-56	-50
Caracas	(-28)	-36	-43	-48	-50	-34
Trinidad	--	(-26)	(-29)	-36	-41	-37
St.José (Costa Rica)	-47	-56	-58	-62	-49	--

[1)] also with Quixeramobim: -39

latitudes. This seems to be indicative for a dominant role of such large-scale features like the Walker circulation.

Table 3

Simultaneous monthly ccf's between RA's of Curaçao and other stations. Significance at the 95% level resp. 99.5% and (90%).

	J	F	M	A	M	J	J	A	S	O	N	D
Quito	33	31	(22)	.	31	24	35	.	.	-24	.	.
Bogotá	.	43	71	(19)	35	32	.	.	30	34	(21)	25
Georgetown	56	48	46	.	.	.	-22	23	23	22	.	35
Caracas	53	29	.	59	43	37	38	(21)	28	23	46	47
Trinidad	61	43	33	.	22	37	32	.	.	.	22	40
Barbados	31	49	.	41	52	32	52	47	23	31	38	33
St.José (Costa Rica)	30	23	.	.	.	.	27	29	34	.	37	35

This coherence is the more remarkable, as here the complete series during a common period of more than 60 years is taken into account.

Certainly a ccf of $r = 0.50$ explains only $r^2 = 0.25$ of the total variance. However, this should be compared with an analysis based on empirical orthogonal functions (EOF's) when only too frequently the first term contributes not more than 20 percent (or even less than 10 percent) of the variance. In such cases an EOF-analysis should always be supplemented by a regionalized (classical) correlation analysis, which allows a better estimate of the reliability - or better the representativeness - of the results. A further check (not applied here) would be an investigation of the stability of the ccf's with time.

C   Hermann Flohn:

ZONAL SURFACE WINDS AND
RAINFALL IN THE EQUATORIAL
PACIFIC AND ATLANTIC

ABSTRACT

Climatological data from the equatorial Pacific (especially from Nauru between 1892 and 1913) indicate a very clear relation between rainfall amount and the zonal component of winds: during dry periods undisturbed easterly winds are dominant, while during wet periods all wind directions occur, as in the vicinity of eddies. This result is confirmed by earlier (1957) simple investigations of a sample of 366 000 marine observations in the Atlantic ($10°S-10°N$).

A further comparison deals with the asymmetric structure of vertical components in the vicinity of the ITCZ in the Atlantic: here the 1957 data set coincides well with the composite analysis of the ITCZ based on the 1974 GATE data (Frank, 1983).

During the last years, several authors - following Wyrtki (1975) - have stressed the role of westerly wind anomalies in the equatorial central Pacific for the evolution of an El Niño-event (Philander, 1981; Rasmussen and Carpenter, 1982). While this refers to time-variations of the prevailing easterly winds, Luther et al. (1983) have stressed the role of real westerly winds, using climatological series between 1949 and 1980 from Ocean Island, Tarawa and Canton Island. In the following tables the dominant role of westerly winds for convective activity, especially for heavy rainfall, shall be investigated. Here we follow suggestions of C.E.P. Brooks and Braby (1921), K. Knoch (1927), Flohn and Hinkelmann (1953) and Ichiye and Petersen (1963).

Looking into the early climatological data from the beginning of the century (Mitt. a.d. Deutsch. Schutzgebieten, 30, 1917), the relation between surface wind distribution and rainfall (resp. cloudiness) can be established beyond doubt. Table 1 gives for three equatorial islands (Malden Island after Knoch, 1927) the resultant wind direction and constancy (both evaluated assuming no variation of wind speed with direction) at local noon for a series of extremely dry and wet months. The original data suggest a more or less bimodal distribution of rainfall frequency and amount. In all three cases during dry months easterly winds are prevailing with a constancy of 71-91 percent - there are only a few observations of winds other than from NE, E or SE. In contrast to this, other wind directions prevail during wet months (with frequent heavy showers or abundant rainfall) at Nauru, where the constancy drops to mere 16 percent. At the two other stations wind from E or NE prevail also during wet months, but other directions occur frequently, probably in connection with cyclonic disturbances in the latitude belts 5-10° on both hemispheres. This corresponds to the coincidence of two large-scale convergence zones: the ITCZ[*)] north of the equator and the SPCZ[*)] south of it. Rasmussen and Carpenter (1982) have indicated that these two zones approach each other (and the equator) during an El Niño-event, due to an increasing convergence of surface winds; it is not clear (and even not very likely) that they really merge.

Table 2 gives the relative frequency of wind directions during wet and dry months at Nauru and Apaiang in detail. For Ocean Island (Table 3) Brooks and Braby (1921) published average rainfall amounts for the 8 wind directions, together with their relative frequency. With winds

---
[*)]
For acronyms see footnote on p. 3.

Table 1

Wind direction during wet and dry months in the equatorial Pacific

		n	$\bar{R}$	$\bar{Rd}$	Cl	dd	Q	Lat.	Long.
Nauru	R > 300 mm	27	429	20.0	62%	328°	16%	0°26'S	166°57'E
	R < 20 mm	24	9,4	3.3	31	95°	71%	Period 1895-1913[1]	
Apaiang	R > 300 mm	9	454	23.0	75%	61°	37%	1°43'N	173°02'E
	R < 30 mm	8	20.5	8.8	50	95°	78%	Period 1906-1909[1]	
Malden	R > 200 mm	15	390	--	--	67°	53%	3°50'S	154°58'W
	R < 10 mm	25	4.0	--	--	89°	91%	Period 1910-1919[1]	

n = number of months
$\bar{R}$ = amount of rainfall per month (mm)
$\bar{Rd}$ = number of rainy days
Cl = cloudiness (originally in tenths)
dd = direction of resultant wind (90°=E; 14^h)
Q = constancy of winds [2]

1) with many gaps
2) assuming wind speed independent of direction

from NE-SE (at 09 a.m.) 4.2 mm/d were observed, in contrast to
14.4 mm/d with winds from NW-SW. This table summarizes the results for
January-June during the years 1905-1909; rainfall frequencies are not

Table 2
Relative frequency of wind directions in units 10^{-3}.

Direction		N	NE	E	SE	S	SW	W	NW	Calm
Nauru	wet	119	104	139	88	50	63	216	142	80
	dry	23	220	402	267	39	21	7	22	3
Apaiang	wet	130	243	222	60	91	28	67	40	118
	dry	64	135	441	308	12	.	2	2	36

given. Orographical causes are probably negligible; the highest hills
at Nauru reach 30 m, at Ocean Island 80 m. SST[*])-variations between wet
and dry months cannot be excluded, but must remain small in this area
west of the date-line, where average SST-values fluctuate around 28 and

Table 3
Frequency n of surface wind direction (09^h) in units of 10^{-3} and
average daily rainfall ($\overline{RR}$) at Ocean Island ($0°54'S$, $169°32'E$)
in units of mm/d

	N	NE	E	SE	S	SW	W	NW	Calm
n	16	146	259	406	15	62	39	30	32
$\overline{RR}$	1.3	6.8	14.8	3.2	16.3	15.8	15.5	10.4	6.4

$29°C$. At Malden Island, situated in the upwelling belt $0-5°S$ east of
the date-line, the situation may be different; no local SST-data are
available. Thus these early data sources demonstrate that undisturbed
easterly winds are correlated with prevailing subsidence and dry - indeed arid - conditions, while heavy rainfall occurs together with individual disturbances, causing winds from all directions. The result
agrees well with the results of Luther et al. (1983, 1984) who show

---
[*]) For acronyms see footnote on p. 3.

(cf. the wind records from Ocean Island 1953-1981 in their Fig. 2) that the El Niño-events are accompanied by individual bursts of westerly winds, without well-defined time relations to the onset of Niño-conditions at the Ecuador-Peru coast. Keen (1982) has described the frequent formation of cross-equatorial tropical cyclone pairs, causing deep westerly winds and excessive rainfall at Canton Island (2°46'S, 171°43'W), together with advection of very warm water (SST $\sim$ 29°C) towards and beyond the date-line.

In a systematic study of more than 366 000 maritime observations in the central equatorial Atlantic (lat. 10°N-10°S, with 1° latitudinal resolution along the shipping-route Europe to South America), Flohn (1957) has investigated the relations between cloudiness and precipitation, indicating vertical motion, and surface wind direction. At least in the belt 0-10°N, SST-variations are small, perhaps negligible; weak upwelling is limited to a belt just south of the equator (Henning and Flohn, 1980). Table 4 gives a summary of all observations. As in the equatorial Pacific, westerly wind components are infrequent (5.9 per-

Table 4
    Rain frequency in % in the equatorial Atlantic (lat. 10°N-10°S) in relation to the zonal component of the surface winds (data mainly 1906-13, 1922-38).

Wind regime	Latitudes 5-10[1])	11-4	Rain frequency with zonal wind components E	O	W
NE-trades	-	5-10°N	6.2	17.1	24.3
ITCZ region	5-10°N	0- 5°N	13.2	20.2	25.5
SE-trades[2])	0- 5°N	-	6.7	12.3	20.4
SE-trades	0-10°S	0-10°S	6.9	17.4	27.0
Whole belt	10°N - 10°S		8.1	18.3	25.1

[1]) 5-10 = May-October

[2]) crossing the equator

cent), but the detailed results for these components are based on about 21 600 observations and thus representative.

In all four (seasonally migrating) climatic zones as well as in the whole belt rain frequencies with westerly wind components are higher by a factor ~3 than with easterly components. This relation is weaker only near the ITCZ, where travelling eddies and/or near-stationary Cb-clusters cause a more equal distribution of wind directions together with rainfall. Splitting the statistics with increasing zonal components, greater westerly components are correlated (nearly independent of latitude, cf. Flohn, 1957, Fig. 4) with higher rain frequencies up to 30-40 percent. Convective activity is also weakly but significantly correlated with the meridional wind components: while equatorward components are accompanied in 8.1 percent of all observations by rainfall, 12.0 percent poleward components (i.e., opposite to an idealized Hadley regime centred at the equator) are observed simultaneously with rainfall (total average 9.9 percent).

Based on the 1° lat. values given by Flohn (1957, Fig. 3 and 7), the correlation coefficient between the average westerly wind component and rainfall frequency amounts to +0.87 (November-April) resp. +0.86 (May-October). These values are derived from 20 independent pairs of data, each averaged over 9000 marine observations before 1939; the significance limit at the 0.1 percent level is 0.68.

As an example, Fig. 1 shows a polar diagram of rainfall frequency in an individual field (0-2°N, 28-32°W) for 16 wind directions. The

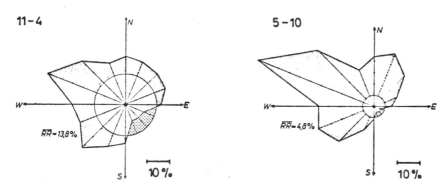

Fig. 1
Rainfall frequency (in %) during southern summer (11-4) and winter (5-10) half-year as a function of wind direction (average values $\overline{RR}$ on the left). Area 0-2°N, 28-32°W.

different sizes of areas (hatched for positive deviations, dotted for negative deviations from average) are caused by the quite different wind frequencies: SE-winds prevail especially during northern summer, when the SE-trades are crossing the equator. The physical reason of this apparently quite general relationship has been ventilated, after some preliminary publications (since 1953), e.g. by Lettau (1956), who discusses the role of the horizontal Coriolis component ($l = 2\Omega\cos\phi$) and especially of turbulent terms.

In their remarkable review papers on the recent Super-Niño-event Cane and Rasmussen-Wallace (1983) describe the sequence of 1982/83 events for ocean and atmosphere separately, without discussing the obvious interactions between them. Here it is not intended to refer in necessary detail to all papers dealing with these interactions. That this event was first observed during May-June 1982 in the central Pacific should not be surprising - in this region the occurrence of westerly winds and high rainfall can happen during each season, necessarily with different consequences. Since the earliest stages of such an event apparently start in the central equatorial Pacific west of the dateline, some hints shall be given to a possible scenario which is independent of the season of the year. This scenario may start with a more or less random evolution of a tropical cyclone (or perhaps two, at lat. 5-10°N and S) in the warm water (~29°C) regions of the western Pacific, causing for a few weeks westerly winds at the equator. In the upper layers of the ocean, this triggers (Philander, 1983) Kelvin waves travelling eastward, transporting warm water towards east and suppressing upwelling. In the atmosphere this causes, in the Indonesian area, mass outflow and divergence, together with a reduction of the rain-producing diurnal circulations between the islands with their high mountains and the sea (Houze). The simultaneous triggering of strong convective activity by the westerly surface winds leads to a displacement of the Indonesian center of atmospheric heating (by the release of latent heat) towards the date-line; an example of this has been described for the 1978-79 Winter MONEX period by Sumi and Murakami (1981). The surface westerlies expand gradually eastwards converging with the easterlies and causing a simultaneous extension of the areas of high SST and of high convective activity towards east. In the case of 1982-83 this evolution spread, during 6-7 months of 1982, across the whole Pacific and arrived, during November, at the Galapagos and soon after at the Ecuador coast, further expanding meridionally. Certainly the chain of events varies substantially with seasons - not each period of warm water and high rainfall in the central Pacific can be followed by a (seasonally fixed) El Niño at the Ecuador-Peru coast. Table 5 (Behrend,

1983) gives ten 12-months-periods with highest and lowest precipitation at Nauru. Both samples are virtually independent of season; this does not recommend a composite analysis on an annual base.

It should also be mentioned that some details of the structure of the ITCZ-region have been demonstrated earlier (Flohn, 1957) which

Table 5

Rainfall amount (mm) for the ten most humid and most arid periods at Nauru (1892-1972), each for 12 consecutive months.

Humid periods	
7/1911 - 6/1912	4600
5/1918 - 4/1919	5047
7/1925 - 6/1926	4133
12/1929 - 11/1930	4752
12/1939 - 11/1940	4390
12/1940 - 11/1941	4132
3/1946 - 2/1947	3906
2/1953 - 1/1954	4360
5/1957 - 4/1958	3940
1/1972 - 12/1972	3979
Average	4324

Arid periods	
6/1893 - 5/1894	507
5/1898 - 4/1899	342
6/1903 - 5/1904	511
10/1909 - 9/1910	395
4/1916 - 3/1917	95
12/1933 - 11/1934	547
3/1938 - 2/1939	279
12/1949 - 11/1950	217
2/1955 - 1/1956	522
6/1970 - 5/1971	374
Average	379

have been confirmed by a careful recent composite analysis with the GATE data (Frank, 1983). It had been shown that the belt of maximum convergence, uplift and rainfall does not coincide with the earlier, simple definition of the ITCZ as a line (or belt) at which the trades from both hemispheres merge, with opposite meridional components (v). The position of the belt of maximum convergence is systematically situated southward of the line defined by v = o: while Frank gives a difference of $2°$ lat., the 1957 paper (averaging over many years and all seasons) obtains $2-4°$ lat. Both papers show that the ITCZ region also coincides, in lower layers, with a near-vertical plane separating negative and positive relative vorticity (Flohn, 1957, Fig. 12; Frank, 1983, Fig. 8); the absolute values are in the same order of magnitude ($2-5 \times 10^{-6} s^{-1}$). However, Frank does not mention the remarkable fact that in the belt between the ITCZ - which varies in the GATE data around lat. 6-9°N - and the equator negative (i.e., anticyclonic) vorticity is observed, together with converging low-level winds and a strong vertical uplift. In the 1957 paper this surprising feature is shown to be a climatological fact, tentatively interpreted as caused by a continuous vorticity advection with the prevailing southerly winds from the southern hemisphere, where negative vorticity is cyclonic.

The average vertical component of motion in the central area of the ITCZ is calculated by Frank to about -180 hPa/d (equivalent to +2.8 cm/s) at the 500 hPa-level; a sample of non-simultaneous pilot-balloon ascents (with their inherent errors) from earlier expeditions yielded in 1957 an estimate of 1.0-1.4 cm/s at 4-5 km. While Frank calculated the rainfall in the ITCZ belt to 18 resp. 20 mm/d, the 1957 paper gave an estimate of 10 mm/d or 300 mm/month, with an asymmetric pattern at both sides of the ITCZ defined with v = o, just as in the Sahel-Sudan belt of Africa (Flohn, 1966). Using this definition, Frank's data yield at 500 hPa an average vertical uplift of -48 hPa/d north of the ITCZ, of -137 hPa/d in a 5° belt south of it (including the band of maximum convergence) and subsidence of +40 hPa/d at a distance of 6-7° lat. south of the ITCZ, i.e., immediately north (0-2° lat.) of the equator. Westerly surface winds prevail in the whole area but are weakly divergent at these latter bands. This latter result can be understood, taking into account the occurrence of upwelling cold water along the equator, which spreads to both sides and causes a belt of very small rainfall up to lat. 2°N. This coincides with the satellite-derived rainfall estimates during GATE (Woodley et al., 1980) and the climatological meridional cross-section (Flohn, 1957, Fig. 7); in the latter representative data source both rain frequency and surface wind convergence reach their minimum at lat. 0-1°N, their maximum at 6-7°N. Between div$\mathcal{V}$ and rain frequency R a significant ccf of -0.73 (in northern winter -0.87) has been found from 18 pairs (Flohn, 1957, Fig. 8); the (annual) regression equation is R = 7.5 div $\mathcal{V}$ (R in %, div$\mathcal{V}$ in $10^{-6}s^{-1}$). These remarkable conformities demonstrate that the results of the 1974 GATE experiment are truly representative, as well as the 1957 derivations, at least for the Atlantic.

D   Hans-Peter Junk:

NAURU RAINFALL 1893 - 1977;
A STANDARD COMPOSITE RECORD

ABSTRACT

The monthly rainfall in Nauru ($1°S$, $167°E$) for the years 1893-1977 is reported in different non-overlapping data sets which show various gaps. By comparison with neighboured stations the gaps (about 14% of total record length) are objectively closed. The completed record consisting of 1008 monthly rainfall sums in the unit 1/10 mm is presented as a basis for further research.

The original data for the rainfall series Nauru were taken from the
following three sources:

1) August 1892 - December 1913:
   Mitteilungen aus den deutschen Schutzgebieten, 30, 65-81, 1917;

2) June 1915 - December 1971:
   An Atlas of Pacific Islands Rainfall, Ronald C. Taylor, Hawaii Institute of Geophysics, August 1973, Data Rept. No. 25, HIG-73-9;

3) January 1972 - September 1977:
   Report of monthly and yearly rainfall, Bureau of Meteorology, Australian Weather Service.

Comments to the sources:

for 1) data-unit: mm (given in 1/10 mm).
   For February 1898 a footnote indicates little or no rainfall at Nauru. The value was therefore set at 0 mm.

for 2) data-unit: inch.
   Data were converted to mm (1 inch = 25.4 mm) and rounded to 1/10 mm.

for 3) data-unit: mm (given in full mm).
   The measurements originate from Nauru-Coastal-Radio-Station. The height of the station is 26 m. Single data exist too for 1982 and 1983 (Nauru Airport). Because of the extensive gaps they were not incorporated into the series.

For the reconstruction of the missing data following rainfall series were used:

1) Ocean Island   lat.: $0°52'S$ lon.: $169°35'E$
2) Onotoa   lat.: $1°51'S$ lon.: $175°30'E$
3) Fanning   lat.: $3°55'N$ lon.: $159°23'W$

These series were also taken from Taylor's atlas.
   For each single month regression-equations were calculated to complete the missing data. The following Table 1 shows the correlation coefficients for each month of the above-mentioned three series with

the basic series Nauru. Also correlations with the rainfall series of Malden (lat.: 4°01'S, lon.: 155°01'E) can be found in Table 1. These series begin in March 1890 up to February 1926.

Table 1
Correlation coefficients with basic series Nauru:

Month	Jan	Feb	Mar	Apr	May	Jun	Jul	Aug	Sep	Oct	Nov	Dec
Ocean Island	.73	.80	.87	.85	.84	.81	.86	.75	.81	.90	.68	.77
Onotoa	.51	.50	.68	.69	.83	.74	.57	.79	.74	.48	.40	.49
Fanning	.30	.46	.31	.37	.44	.37	.47	.56	.70	.57	.55	.41
Malden	.26	.61	.36	.40	.10	.29	-.06	.31	.40	.21	.39	.25

The following gaps of the Nauru series were filled:

May 1905
November 1906 - March 1910
January 1914 - May 1915   with Ocean Island
July - December 1920
November 1925 - August 1926
December 1940 - February 1941
October, November 1941

January 1944 - October 1945
January, September 1969   with Onotoa

December 1941 - December 1943
August 1971   with Fanning

The gaps before 1905 could only be closed with long-term monthly means, respectively in one case with linear interpolation. Because of the (in most cases) insignificant correlations of the Nauru series with the only series available before 1905 (Malden), a regression computation was not applicable. The September 1895 gap was filled through linear interpolation between August and October, since during this time there was obviously a dry period and consequently a monthly mean would probably be too high. For the months January - July 1901 and March,

Table 2

Reconstructed rainfall series NAURU: October 1893 - September 1977
(lat.: 0°34'S   lon.: 166°55'E)   Data in units 1/10 mm

Year	Jan	Feb	Mar	Apr	May	Jun	Jul	Aug	Sep	Oct	Nov	Dec
1893										118	72	1293
1894	81	43	18	38	118	1706	627	1127	454	160	1285	577
1895	2770	3187	1335	65	946	538	230	472	717	961	2236	3064
1896	3032	1985	1794	1370	1855	1021	2866	2157	2532	3409	1570	5595
1897	2716	2985	1417	753	849	1046	1140	188	693	727	1578	328
1898	025	0	106	577	160	215	445	850	195	165	125	610
1899	175	260	10	180	510	2870	1900	3020	1045	1300	2680	4175
1900	1130	7530	2815	3520	3645	2330	1105	3630	950	1070	3315	2625
1901	2651	2449	1940	1607	1125	1187	1548	723	505	41	265	1831
1902	1281	1298	1940	1607	3869	2634	4334	1802	4178	1035	4059	2990
1903	2412	3706	989	1759	1697	117	413	361	235	7	636	861
1904	762	206	239	23	1245	536	2515	3174	2943	1042	4143	2712
1905	4321	3986	1337	3934	2608	2475	4342	1881	1916	3636	1620	4960
1906	4772	2548	1759	556	487	826	441	341	182	354	3776	1948
1907	2401	1294	2140	2296	393	860	974	673	573	2239	2510	1792
1908	5231	2763	464	582	442	1900	1023	1450	533	253	565	1093
1909	1455	727	699	576	289	737	1731	1056	872	462	754	711
1910	1209	941	471	204	1132	410	800	275	78	700	56	548
1911	3026	2580	5205	3466	1664	1356	4504	1739	5484	2866	1980	3429
1912	4199	6834	5430	4427	3456	1652	1036	912	1603	3494	1106	7303
1913	2285	2190	427	32	238	273	594	2289	2258	2618	2003	2517
1914	4013	3271	3835	3916	3735	2510	6448	2574	1873	734	2506	3892
1915	3717	2587	2263	3138	4888	922	714	79	226	0	53	348
1916	3218	264	399	51	38	183	203	198	13	74	0	23
1917	94	33	51	732	102	874	574	218	1285	119	114	612
1918	43	348	3825	856	1313	1636	4783	7615	3045	4869	3541	3256
1919	4394	6441	5116	4539	1049	1331	2037	3470	3287	3500	2410	2929
1920	597	6027	4790	1829	1265	2235	1796	800	558	298	585	2300
1921	2720	749	201	224	122	23	803	1140	566	538	945	4514
1922	3769	2075	173	879	856	1158	3076	2769	577	91	1651	300
1923	1057	876	2888	1732	318	1666	1727	3571	1890	2215	6731	3978
1924	4348	1532	1593	36	605	615	833	836	302	584	145	33
1925	495	1125	84	5	472	914	1077	2438	1791	4742	1521	3251
1926	3082	3562	5047	4022	2761	1694	1218	2177	759	249	1948	320
1927	183	74	38	406	376	980	1572	1049	716	114	2327	3955
1928	5491	1755	254	815	122	145	770	366	1255	335	1389	4829
1929	3467	3937	218	224	554	737	2522	2484	1097	993	2322	5286
1930	4653	4481	5611	1364	818	2614	2804	6947	6881	2720	3350	3635
1931	4544	1849	4460	1204	599	1826	376	1712	191	940	323	1732
1932	8606	5565	4382	2276	1024	1008	2096	1666	800	0	23	56
1933	3673	1151	810	1692	950	1623	2614	942	61	572	178	0
1934	315	46	343	168	378	1626	1290	881	180	124	127	3533
1935	3592	2228	465	99	998	886	1303	130	909	1605	757	5029
1936	3828	2644	3340	1105	292	330	450	1077	490	64	3924	3106
1937	2764	4031	1623	127	333	450	757	1260	975	345	117	127
1938	2070	1207	137	13	23	254	368	173	33	0	411	673
1939	373	338	333	91	701	1234	1824	2807	2085	523	3302	4478
1940	2880	3736	3828	5187	3076	3048	3566	2631	3160	4089	4496	3909
1941	3111	2246	3876	4486	2479	3487	3378	2113	3830	2814	2884	2415
1942	3787	2075	2727	1977	936	880	808	1452	263	699	878	1898
1943	2353	2029	1559	1612	1018	1083	1949	1938	395	784	1219	2908
1944	3286	1884	1778	1803	197	1551	1207	1171	966	807	1751	1672
1945	2117	1852	1118	1332	1302	2623	1534	1006	587	752	396	1290

Table 2 (continued)

Year	Jan	Feb	Mar	Apr	May	Jun	Jul	Aug	Sep	Oct	Nov	Dec
1946	2248	2578	3332	3655	3094	3459	3967	1943	2189	4585	3950	2278
1947	3686	2921	91	533	241	1953	1288	714	124	112	1854	3332
1948	2583	3868	6436	4808	4272	2997	1425	1110	91	366	2652	3053
1949	5875	1133	3899	1687	439	406	587	582	249	498	300	295
1950	18	356	20	56	71	124	175	508	196	102	254	904
1951	3068	231	2156	2667	4069	2774	2342	4056	1067	2852	1415	5972
1952	2576	1339	1036	1171	1062	1011	1862	2029	1549	351	282	3637
1953	4100	6231	3861	4161	4432	1537	2390	4267	1303	4105	1885	5283
1954	3673	1572	2205	1567	262	221	198	711	815	15	13	155
1955	4864	229	947	457	439	328	193	284	782	18	104	409
1956	1039	1214	780	643	526	69	198	437	160	399	823	1377
1957	1572	4150	2748	1763	2217	2593	2731	2101	2416	3566	4140	4171
1958	2565	5575	2644	4686	1981	2134	3162	1229	627	442	1420	1542
1959	3846	2296	1533	986	653	940	277	208	114	305	1615	4407
1960	5763	2451	886	1902	1532	71	1732	467	805	983	1402	3559
1961	2377	2187	3995	2667	1397	1420	1478	2469	582	495	1775	592
1962	224	279	323	683	627	531	1041	1044	803	15	3444	582
1963	343	132	259	889	683	1001	516	1488	2715	3553	1316	4516
1964	3058	5679	1255	33	71	38	244	752	183	70	604	2822
1965	3998	3950	1341	3155	1040	1920	3444	3871	3249	3851	3564	3246
1966	3317	4412	2687	3015	1585	1138	2148	394	99	1057	218	1328
1967	635	597	137	4064	465	173	815	437	488	107	165	5250
1968	4549	3736	170	511	36	137	1405	152	188	64	511	1458
1969	3729	3739	5474	3330	996	536	315	351	686	770	871	3731
1970	2604	6406	3805	4735	714	1046	234	257	30	84	224	0
1971	130	43	10	1151	536	1080	719	2090	175	284	274	1803
1972	3590	1750	250	1720	2930	3000	3430	2870	5260	2200	1830	4820
1973	3350	4320	1500	450	240	1660	120	120	150	40	10	150
1974	0	10	100	1180	1250	580	730	1290	30	60	700	3550
1975	7810	2150	1990	700	750	1140	960	580	730	30	20	170
1976	50	2010	5230	3200	1980	2130	2580	2850	4250	2750	1790	4080
1977	4940	2680	5160	3460	1750	750	5570	2100	2800			

April 1902 monthly means were inserted. The reconstruction series begins not before October 1893. Because of the big gap from December 1892 - September 1893, which only could have been filled with monthly means, the four original values from August - November 1892 were not used for the Nauru series.

Now the reconstructed Nauru rainfall series covers 1008 monthly data from October 1893 - September 1977 (84 years). 140 gaps were closed. So 86.1% of the series are original data. A first investigation using Blackman & Tukey power spectrum analysis shows no significant change in the spectrum compared with the results of Fleer (1975).

E   Karl-Heinz Weber and Hermann Flohn:

OCEANIC UPWELLING AND AIR-SEA-EXCHANGE
OF CARBON DIOXIDE AND WATER VAPOR AS A
KEY FOR LARGE-SCALE CLIMATIC CHANGE?

ABSTRACT

In the upwelling area of the equatorial Pacific (long. 157°W), warm/wet and cold/dry phases of 6 months each were selected. During the cold/dry phase the increase of the atmospheric CO_2 content is significantly lower than during warm/wet (El Niño) phases; the difference is equivalent to 2-3 Gt carbon annually. This is due to the enhanced net biological productivity of the nutrient-rich cool water from below, which consumes about 1-4 Gt carbon annually both from air and water. Simultaneously the H_2O exchange through the air-sea interface (evaporation) drops in upwelling areas to 20-25% of its normal value in tropical oceans.

Since both gases are responsible for the natural greenhouse effect, these fluctuations control global temperatures and atmospheric circulation. Changes in the frequency or intensity of upwelling/downwelling can be triggered by a hemispheric feedback process. Such feedback processes are assumed to be responsible for the observed strong fluctuations of CO_2 and temperature under purely natural conditions, e.g., during and since the last glaciation. Some thoughts on the geophysical mechanism of climatic fluctuations on the 10^2-10^3 year scale are presented.

E.1 INTRODUCTION

Since their beginning in 1958, records of the atmospheric content of CO_2 show an uninterrupted increase, due to the burning of fossil fuel and deforestation. However, the interannual variability of the growth rate of the atmospheric CO_2 is high and quite independent of the consumption rate of fossil fuel. The oceans which may act as regional sinks as well as sources of CO_2 obviously play an important role in this context. Of special interest are the upwelling areas of the equatorial Pacific and Atlantic Oceans, because of the biological consumption of CO_2 in these nutrient-rich waters which must be considered in addition to the physical and chemical processes at the air-sea interface (Newell, Navato and Hsiung, 1978; Baes, 1982, 1983). They also control the evaporation of tropical oceans and thus the global water budget (Flohn, 1983a).

The possible role of such changes of the two most important greenhouse gases for the natural (i.e., non-anthropogenic) climatic changes of the past is discussed. New evidence (Oeschger-Stauffer, 1983) indicates a higher frequency of "abrupt" climate changes on a time-scale of 100 years, tending to recur at intervals of a few millennia. These abrupt changes are accompanied by CO_2-content variations at the same scale (about 70 ppm/100 years) as the present anthropogenic changes (Elliott, 1984); a geophysical interpretation shall be proposed.

E.2 SST - CO_2 CORRELATIONS

Several studies gave evidence for a relationship of the growth rate of atmospheric CO_2 and the SO[*], as indicated by the SOI[*] and the SST[*] of the equatorial Pacific Ocean. Among others, Newell et al. (1978), Angell (1981) and Schnell et al. (1981) found positive correlations of CO_2 increase at several stations with equatorial Pacific SST, whereas Bacastow and Keeling (1981) showed a negative connection with SOI, which itself is negatively correlated with SST in the eastern equatorial Pacific.

Using local data our own studies present additional evidence for this connection. Fig. 1 shows the time-series of monthly anomalies of SST in 1958-78 at Christmas Island (1.5°N, 157°W) (a) and its high serial correlation (b), significant up to seven months. The anomalies ΔT_s are defined as the departure of monthly values from their long-term means. This time-series is only quasi-homogeneous, due to the fact that some gaps (together 14 months) in the data set given by Seckel and Young (1977) are filled by linear regression with Canton Islands' SST. Furthermore, the series is extended to the years 1975-78 by linear regression with Fanning Island's SST (Vitousek, 1979). Despite these shortcomings, the record shows fairly good qualitative agreement with Weare's (1984) SST-series in the area 0-10°S/80-135°W. That T_s-anomalies exhibit a marked coherence in the eastern and central Pacific has been shown by Wright (1977).

Fig. 2 shows the correlation of moving 6-months-averages of the SST-anomalies with the increase of atmospheric content of CO_2, $\Delta CO_2/\Delta t$ (Δt = 6 months) at Mauna Loa two months later, after removal of the annual cycle of the CO_2-series. The correlation coefficient $r = 0.31$ is weak, but significant with respect to the 95% limit.

The relationship becomes more evident if one looks at strong SST-anomalies only, representing phases of strong upwelling or its absence. Therefore, 12 warm phases and 10 cold phases (not overlapping) with an average SST-anomaly $|\Delta T_s| > 0.8°$ persisting during 6 months were selected and the corresponding CO_2-increase (with a lag of two months) calculated. During the warm periods with a mean anomaly of +1.1°C the average CO_2-increase was +1.06 ppm per 6 months, whereas during the cold phases with a mean anomaly of -1.2°C its value was only +0.38 ppm.

---

[*] For acronyms see footnote on p. 3.

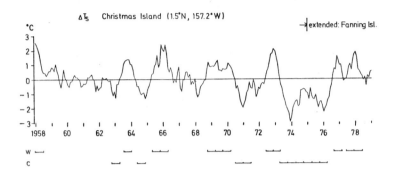

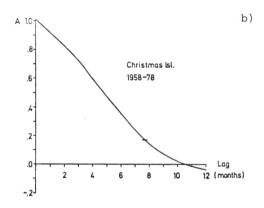

Fig. 1

Time series of monthly SST-anomalies at Christmas Island in 1958-78 (a) and its autocorrelation (b). w and c denote selected six-months-periods with an average anomaly $|\Delta T_s| \geq .8°C$.

The random probability of the difference of these averages is below 5% (Student's t-test applied).

Another variable, strongly correlated with SST, is rainfall in the equatorial Pacific. The correlation coefficient between monthly SST-

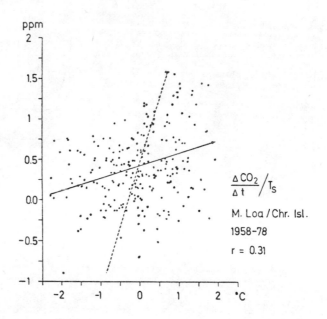

Fig. 2

Correlation between monthly SST-anomalies at Christmas Island and the increase of CO_2 at Mauna Loa during six months; lag = 2 months (CO_2 after SST).

------ , ———— : lines of regression

anomaly at Christmas Island and the Line Islands' Rainfall Index (Meissner, 1976) is r = 0.7 (Behrend, 1983). A selection of 10 wet and 9 dry phases in the years 1958-74 (Table 1) shows a similar difference between the two corresponding CO_2-growth rates, also with random probability of less than 5%.

Table 1 shows that the interannual differences of CO_2-exchange through the air-sea interface are of the same order of magnitude as the annual input of fossil CO_2 into the air ($\sim$2.5 Gt C; 1 Gt = 10^{12} kg).

The net biological productivity of upwelling regions has been measured in coastal regions (Weichert, 1980; Lenz, 1981) to 1.5-3.5 g C/m^2d, in extreme cases up to 10 g C/m^2d. For an upper 20 m-layer in the equatorial Pacific ($5°S-5°N$), Broecker and Peng (1982, p. 398) give a verti-

Table 1
Increase of atmospheric CO_2 (annual cycle removed) following six-months-averages of warm/cold anomalies of Christmas Island and wet/dry anomalies of the Line Islands Rainfall Index with a lag of two months.

	T_s(°C) Christmas Isl.		Rainfall Index (%) Line Isl.	
	warm	cold	wet	dry
average anomaly	1.1	-1.2	75	25
σ	.37	.40	8.5	8.5
$\frac{\Delta CO_2}{\Delta t}\left[\frac{ppm}{6\,months}\right]$	1.06	.38	.67	.10
σ	.48	.27	.46	.46
equivalent to an annual increase of	1.36 ppm or 2.9 GtC/a		1.14 ppm or 2.4 Gt C/a	
number of cases	12	10	10	9

cally averaged net productivity of about 0.2 g C/m^2d, which yields, for an area of $10°$ lat. x $100°$ long. (12.3 x 10^6 km^2), about 2.5 x 10^{12} g C/d or 0.9 Gt C/year; this may well represent a lower limit. Assuming a total upwelling area of 10 x 10^6 km^2 (less than 3% of the total ocean surface), seasonal upwelling during 6 months and an average net productivity of 2 g C/m^2d, one obtains a global annual productivity near 3.6 Gt C by photosynthesis which is in the right order of magnitude. Contributing to this productivity are two sources of CO_2: the super-saturated ocean water and the overlying atmosphere.

E.3 SST - H_2O CORRELATIONS

Possible climatic impacts of a CO_2-increase are amplified (weakened) if there exist positive (negative) feedback mechanisms in the climatic system (Kellogg, 1983). A positive feedback process, for example, is established by a higher content of water vapor in the atmosphere as a consequence of a CO_2-induced global warming; the interaction of such processes has been described by Ramanathan (1981).

In the following studies data from three upwelling areas of the tropical Atlantic Ocean were evaluated to show the connection of evaporation and SST (Fig. 3).

The data are taken from the Marine Climatological Summaries of the years 1961-70, published by the Seewetteramt Hamburg of the Deutscher

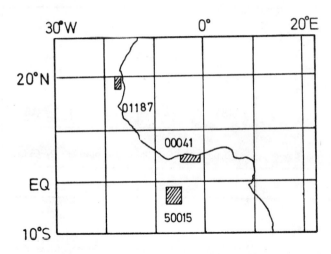

Fig. 3
    Position of three coastal (01187, 00041) and one equatorial (50015) upwelling fields of the tropical Atlantic Ocean.

Wetterdienst. In the field 01187 off the north-west coast of Africa upwelling occurs from January to June, whereas in the field 00041 (Gulf of Guinea) upwelling exists from July to October and south of the equator (50015) from June to October. Field 50015 is of special interest because of its distance far off the continent, where with steady southerly winds only maritime air masses are advected during the upwelling season.

Fig. 4 shows the annual trend of sea surface temperature T_s, sea-air temperature difference T_s-T_a, the difference q_s-q_a of the specific humidities at the sea surface, assuming saturation, and in the air (≈ 10 m), and evaporation or latent heat LE. Evaporation E is calculated

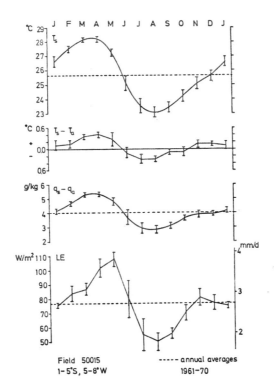

Fig. 4

Seasonal variation of sea-surface temperature T_s (°C), sea-air temperature difference T_s-T_a (°C), difference of specific humidities at sea-surface and at deck level q_s-q_a (g/kg) and latent heat flux LE (W/m^2) or evaporation E (mm/d).

with the well-known bulk equation, depending linearly on q_s-q_a and on wind speed. The decrease of q_s-q_a and LE due to lower water temperatures and higher stability ($T_s-T_a < 0$) is easily recognized. Of course a dependence of evaporation or latent heat flux on T_s is provided mathematically by applying the bulk equation (q_s depends on T_s via the

Clausius-Clapeyron equation). But the relationship is not trivial since LE depends also on the specific humidity of the air and on wind speed.

However, in most areas presented here the influence of wind speed on the annual range of latent heat flux LE is of minor importance; in field 50015 it varies only between 4.4 m/s in March and 6.0 m/s in May (not shown here). The regression of anomalies of T_s and q_s-q_a is shown in Fig. 5. The correlation coefficient is $r = 0.77$; with LE instead of q_s-q_a it drops to 0.66. In field 00041 the correlation is similar:

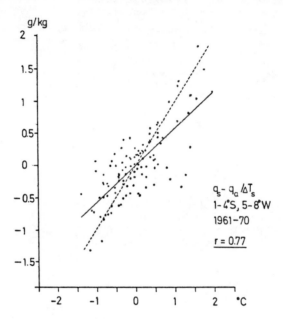

Fig. 5

Correlation of monthly anomalies of SST (T_s) and humidity difference (q_s-q_a) in field 50015.

------ , ———: lines of regression

$r = 0.75$ for T_s versus q_s-q_a and $r = 0.72$ for T_s versus LE. In each case monthly anomalies are correlated.

In field 01187 (coast of NW-Africa) $r = 0.60$ between SST and humidity difference but only 0.20 between SST and evaporation, due to the fact that in that region SST-anomalies are negatively correlated with

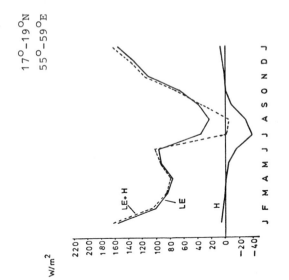
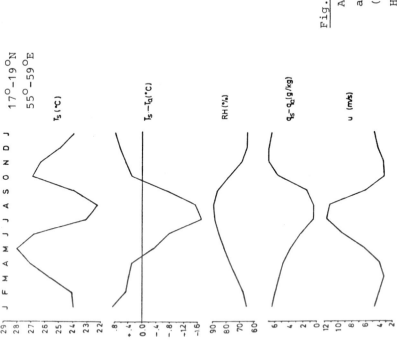

Fig. 6

As Fig. 4, here for the western Arabian Sea; in addition annual trends of relative humidity RH (%), wind speed u (m/s) and sensible heat flux H (W/m^2) are presented. Period: 1956-76.

wind speed. The annual trend also shows a suppression in the upwelling months (January-June) with an average evaporation rate of 2.5 mm/d compared with 3.7 mm/d in the months July to December, while winds are weaker than in the first half of the year. Even lower evaporation values down to 0.8 mm/d have been found by Henning and Flohn (1980) using monthly data for 1953-65 along the shipping route from Europe to South Africa. For the Galapagos region of the Pacific (0-3°S, 85-90°W) Trempel (1978) has also obtained an evaporation drop to 0.8 mm/d in November-April compared to 3-3.5 mm/d in two adjacent fields farther north and south.

Weare (1984) has shown that in the eastern equatorial Pacific SST-anomalies contribute ca. 37% to interannual LE-variations whereas wind anomalies explain 22-36% of these changes with decreasing influence towards coastal waters. In other parts of the Pacific between 30°S and 30°N wind speed is dominant.

Another area with seasonal upwelling exists in the Arabian Sea at the coast of Arabia between ∼15 and 20°N and at the Somali coast (∼5-12°N, 50-55°E). Here extremely strong winds parallel to the coast cause upwelling during the SW-monsoon period from June to September. East of the Somali coast wind speeds up to 16 m/s in July lead to an evaporation maximum despite the stable stratification ($T_s - T_a \approx -0.8°$) (Weber, 1981), whereas at the Arabian coast (Fig. 6) with still high wind speeds (∼12 m/s) but a stronger stability ($T_s - T_a \approx -1.4°$) evaporation is reduced to a minimum of about 1 mm/d in August. The fluxes of latent and sensible heat nearly compensate each other in July/August in that region.

E.4 PARALLEL VARIATIONS OF ATMOSPHERIC CO_2 AND H_2O

The study of various ocean areas with seasonal upwelling shows a strong dependence of evaporation on sea-surface temperature. This is also to be seen by correlations of monthly anomalies of T_s and of the stability of T_s-T_a with the humidity difference q_s-q_a and in almost all cases with the evaporation itself. This evidence justifies the conclusion that oceanic upwelling simultaneously causes a (relative) drop of atmospheric CO_2-content (via biological productivity) and of evaporation (by a reduction of H_2O-exchange through the air-sea interface). Similarly suppression of upwelling - i.e. downwelling in fully developed El Niño-situations - produces simultaneously substantial decreases - or increases - of the CO_2 and H_2O-content of the air in large areas of the Tropics. Since both gases are most efficient absorbers of infrared (terrestrial) radiation, this should have a remarkable effect on the atmospheric heat budget, particularly in tropical areas.

The intensity of upwelling along the equator depends on the curl of the wind stress vector (Yoshida and Mao, 1957); in areas with a fairly homogeneous zonal wind field easterly (westerly) winds cause Ekman divergence (convergence) and thus upwelling (downwelling) (Bjerknes, 1966). In five El Niño-years (1958, 1963, 1965, 1969, 1972) the annual increase of CO_2 (averaged from Mauna Loa and South Pole) is 1.04 ppm, in contrast to 5 years with prevailing cold water (1960, 1964, 1967, 1971 and 1974) with an increase of 0.57 mm/a only. The El Niño-years 1976 and 1982 confirm the marked increase.

Similarly, Pan and Oort (1983) have shown that the water vapor content of the tropical troposphere has substantially increased during northern winter, in the cases of warm SST in the key area of the equatorial Pacific. The global water vapor content as averaged over the whole mass of the atmosphere responds similarly (cf. Oort, 1983, Fig. 44) if one takes into account a 4-6 months-lag, similar as for the three-dimensionally averaged temperature.

E.5 ABRUPT CLIMATIC CHANGES UNDER NATURAL CONDITIONS; THE ROLE OF CO_2

The discovery of surprisingly large variations of the CO_2-content of air bubbles enclosed many millennia ago in the ice-sheets of Greenland and Antarctica (Berner et al., 1980, Delmas et al., 1980), i.e. under purely natural conditions, has opened a new dimension to the CO_2-climate problem. Disregarding many fascinating facets, we draw our attention here solely to the (likewise new) problem of abrupt climatic changes (Flohn, 1979, 1983a, b), at a time-scale of centuries or even less. A selection of cases (Flohn, 1984) is given in Table 2; but in the meanwhile Oeschger and Stauffer (1983) have demonstrated (Fig. 10) that

Table 2

Abrupt palaeoclimatic events (time-scale ~100 a).

ΔT			Time BP
a) W	Transition Y. Dryas - Preboreal		10.2 ka
b) C	"	Alleröd - Y. Dryas	10.8
c) ?	Onset Moist Period (lat. 15-35°N)		12
d) W	Transition O. Dryas - Bølling		12.8
e) C	"	Stages 5a/4	73
f) C	"	" 5c/5b	95
g) C	"	" 5e/5d	115
h) C	Cold Episode Holstein Interglacial Stages 7, 9?		190?
i) C	" "	Stage 19 (Matuyama - Brunhes)	700

many more cases do exist. Their recurrence time apparently is in the order of a few millennia - similar to the recurrence of cool episodes - as, e.g., the "Little Ice Age" - during the Holocene (Denton-Karlén, 1973; Karlén, 1976, 1979; Gamper-Suter, 1982).

Since after the discovery the first papers indicated only one strong CO_2-increase at the transition between Late Glacial and Holocene, about 10 000 years ago, early hypotheses were presented, which finally ended in a deadlock. Broecker (1982) interpreted this rise as caused by marine sedimentation in slowly submerging shelves during the glacio-eustatic rise of the sea-level. But this mechanism is much too

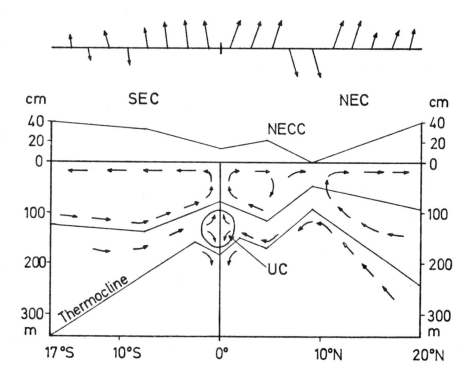

Fig. 7

Oceanographic conditions in the equatorial Pacific between 150 and 160°W, showing surface currents (above), dynamical height (cm), thermal structure and meridional circulation (after Wyrtki and Kilonsky, 1984). SEC (NEC) = Southern (Northern) Equatorial Current, NECC = Northern Equatorial Counter Current, UC = Undercurrent.

slow (Oeschger et al., 1982). On the base of increasing evidence for abrupt climatic changes on the 10^2 year-scale, it has been suggested that substantial changes in the frequency, duration and intensity of upwelling or downwelling - the "Super-Niño" of 1982/83 may serve as a drastic example - are able to reduce or to increase the atmospheric CO_2-content by values of 60-80 ppm (as observed) during the course of a century or so. This could be caused, e.g., by 100 half-yearly events as described in Table 1. More likely, however, is a cluster of much more intense (or prolonged) events; such a cluster may occur at a rate of 10^{-3} per year. A series of large volcanic eruptions (as, e.g., between 1810 and 1840) or a lull (as between 1912 and 1948) could trigger, on a smaller scale, such a hemispheric feedback mechanism (Flohn, 1982, 1983; Kellogg, 1983) between atmospheric and oceanic circulation.

Oceanographic investigations (Wyrtki, 1981, Wyrtki and Eldin, 1982) indicate that the oceanic exchange processes are limited to a quite narrow zone and to shallow depth. How far they are related to the large-scale circulation of the whole ocean, with the main sources of deep and intermediate water masses in the arctic North Atlantic and along the Antarctic (cf. Broecker-Peng), remains an open question. The transverse circulation in the equatorial Pacific (Fig. 7) has been quantitatively described by Wyrtki (1981; cf. also Wyrtki-Kilonsky, 1984) on the base of the Shuttle data between Hawaii and Tahiti. Equatorial upwelling is limited to a narrow zone on both flanks of the equator, just above the equatorial undercurrent; its average intensity is in the order of 1 m/d, yielding over 10^4 kms zonally and 400 kms meridionally a flow of 40-50 Sverdrup (1 Sv = 10^6 m^3/s). It is maintained, at and below the level of the undercurrent, by an asymmetric convergent mass flow of the same magnitude; averaged over the layer 50-200 m depth this is equivalent to a meridional flow of 7 cm/s, the major portion arriving from south. Fig. 7 indicates that this convergent flow is situated within the main thermocline, which can be defined here as the layer enclosed by the isotherms $26°C$ and $14°C$. Fig. 8 shows the strong gradient of nutrients, especially of nitrogen, in the thermocline. Both phosphorus and nitrogen indicate a broad upward transport at the belt 0-6°S in the uppermost 100 m. A narrow upward directed tongue of high nitrogen content at the equator, above 100 m, coincides with a depletion of O_2; both indicate the dynamical role of the narrow undercurrent (see U in Fig. 9) fed by meridional convergence and maintaining vertical divergence (upward above, downward below, cf. Fig. 7). The marked asymmetry to the equator - which is also observed in the At-

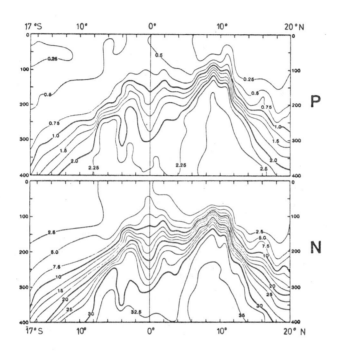

Fig. 8

Mean distribution of phosphate (P) and nitrate (N) in µM/kg between Hawaii and Tahiti and from the sea-surface to 400 m (after Wyrtki and Kilonsky, 1984).

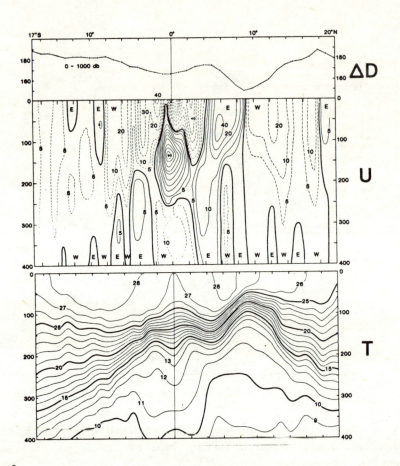

Fig. 9

Mean distribution of dynamic height ΔD relative to 1000 db in dynamic cm, zonal geostrophic flow u in cm/s and temperature T in °C between Hawaii and Tahiti and from the sea-surface to 400 m (after Wyrtki and Kilonsky, 1984). Ocean currents are denoted towards E (or W).

lantic, but not in the Indian Ocean with its seasonal reversal of winds and currents - is shown in Fig. 9. The baroclinic NECC[*] towards E in lat. 4-8°N - with a thermal gradient of 12°C/500 km at a depth of only 120 m - creates its own transverse circulation and limits the meridional transport of nutrients to the latitude belt 0-4°N. In contrast to this, a southern countercurrent is hardly distinguishable, only marginally developed as an ephemeral feature around 10°S. Thus the meridional transport of nutrients originates mostly from the southern hemisphere, where its vertical component is stronger.

These empirical facts convincingly indicate that the oceanic circulation, which maintains equatorial upwelling, is essentially limited to the belt 10°S-4°N and to a depth not much below 200 m. If we assume for both Atlantic and Pacific a total area of 14° lat. and 140° long., $\sim 25 \times 10^6$ km^2, and a depth of 200 m, the volume involved is about 5×10^6 km^3 or 0.4% of the total volume of the ocean. There is no obvious correlation between this shallow and narrow transverse circulation in equatorial latitudes and the great circulation produced by the bottom-water production in the subarctic Atlantic and along the Antarctic Ice, which comprehend more or less the whole global volume of the oceans.

Both circulations are of a different spatial scale - they can be expected to work also on a different time-scale. With the exciting results of the physical laboratories in Bern and Grenoble (Lorius and Raynaud, 1983; cf. the most recent review given by Elliott in late June 1984), together with data from other sources, the attempts to use slow palaeo-oceanographic effects to explain these natural variations of the CO_2-content have become obsolete (Broecker, 1984). The observed changes on the interannual scale - the distance between El Niño-events varies between 2 and 7 (or even more) years - may serve as a rational geophysical model, which can produce natural global CO_2-variations as related to the coupling between atmospheric and oceanic circulations in the tropics. These exchange processes are limited to the uppermost 200 m-layer of the ocean: instead of "changes in the deep-sea ventilation rate" (Broecker, 1984; cf. also Siegenthaler and Wenk, 1984) we deal here with changes in the nutrient content in the upper mixing layer and consequently in the net bio-productivity, which are responsible for an essential part of the CO_2-exchange between ocean and atmosphere. In-

---

[*] For acronyms see footnote on p. 3.

deed, the role of the oceans in the CO_2-budget cannot be correctly understood if the biological processes are underrated. Their role has recently been pointed out, among others, by Baes (1983), Fiadeiro (1983) and Keeling et al. (1984); cf. also Newell and Hsiung (1984).

E.6 A GEOPHYSICAL INTERPRETATION OF LARGE-SCALE CLIMATIC CHANGES ON THE 10^2-10^3-YEAR SCALE

Flohn (1982, 1983a) and Kellogg (1983) have described a positive feedback mechanism, which starts from the observed correspondence between zonal winds and the above-mentioned vertical motion in the oceanic Ekman layer along the equator (Wyrtki, 1975, Reiter, 1978). This mechanism (Table 3) may have contributed substantially to the simultaneity of tropical aridity and low CO_2-content (180-220 ppm) during the peak

Table 3

Natural hemispheric climatic feedback.

	cooling	warming
Initiation[1] (polar regions)		
Meridional ΔT		
Intensity of westerlies	$\rightarrow$	$\rightarrow$
Lat. subtropical high (ϕ_{STA})	equator	pole
Intensity of trades (Hadley cell)	strong	weak
Equatorial Oceans[2]	upwelling	downwelling
SST Equatorial Oceans	cold	warm
Atmospheric content CO_2 + H_2O	$\rightarrow$	$\rightarrow$
resulting in further	cooling	warming

[1] triggered, e.g., by a cluster (an absence) of volcanic eruptions
[2] except Indian Ocean

and decay of the last glacial about 20-14 ka ago, as well as to the occurrence of a humid climatic phase at the tropical continents together with high ocean temperatures and high CO_2-content (300-320 ppm) during the Holocene warm epoch about 10-8 ka ago (Oeschger et al., 1983; Flohn, 1983). Similarly it is coherent with the quasi-simultaneous occurrence of these epochs at both poles, in contrast to an alternation as expected from orbital elements (precession). In addition to this, this feedback mechanism presents, if correct, a reasonable approach to the enigma of abrupt climatic changes on a time-scale in the order of a century.

This problem of abrupt climatic change (Flohn, 1984) has now greatly benefitted from recent studies at the Greenland ice core Dye 3

(Dansgaard et al., 1982; Oeschger et al., 1983; Oeschger and Stauffer, 1983). Parallel measurements of the $^{18}O/^{16}O$ ratio, as representing ambient temperature, and of the CO_2-content of the air bubbles enclosed in the ice indicate that between about 70 ka (ka = Kiloanno = 1000 years) and 9 ka ago a sequence of "abrupt" changes occurred. They demonstrated well-correlated (within the limits or error) simultaneous changes of temperature and CO_2-concentration. The accuracy of the CO_2-measurements has been thoroughly checked (Barnola et al., 1983) and substantially improved using laser spectroscopy and a crushing system needing not more than 1 cm^3 of ice. Absolute dating of the core has not yet been possible, while intercomparison with the wealth of available evidence gives at least a satisfactory estimate. The relative time-scale, at a depth of 1850-1900 m, is then 1 m corresponding to 300-350 years.

Fig. 10 shows the fluctuations of both parameters at the interval of 1860-1890 m depth (about 27-37 ka BP). Due to the slow metamorphosis of snow into firn and ice, the individual measurements of CO_2 (circles with an accuracy of 6 ppm) represent an average of about 100 years; this limits the time-resolution of an ice core. The complete $\delta^{18}O$ record (Dansgaard et al., 1982) and a comparison between this core and the Camp Century core (Oeschger et al., 1983, Fig. 1; Oeschger-Stauffer, 1983, Fig. 5) - in a distance of nearly 1500 km - indicate that there are frequent climatic fluctuations (not strictly periodic) on a time-scale around 2-3 ka. During the last ice-age - at least between about 40 ka and 10 ka ago - the CO_2-content varied between 180-200 ppm (during cold stadials) and about 250 ppm (during warm interstadials). Oeschger-Stauffer's Fig. 8, probably about 40 ka ago, contains about 30 individual CO_2-measurements per meter depth. It demonstrates clearly, at 1897.6 m depth, that the transition from an average level of 190 ppm up to an average level of 256 ppm, both values valid for about 400 years each, took not more than about 100 years (equivalent to a 32 cm-layer). The synchronous $\delta^{18}O$ difference is 5.2°/oo. A quantitative model including exchange processes through the thermocline, but neglecting details of the coupling processes between atmospheric and oceanic motions and its time-scale, has been presented by Siegenthaler and Wenk (1984).

The measurements of $\delta^{18}O$ and CO_2, as derived from Fig. 10 (Oeschger-Stauffer, 1983), are not precisely synchronous, since adjacent parts of the core had to be used for the different measurements. Nevertheless, it is possible to compare each of the turning points of the

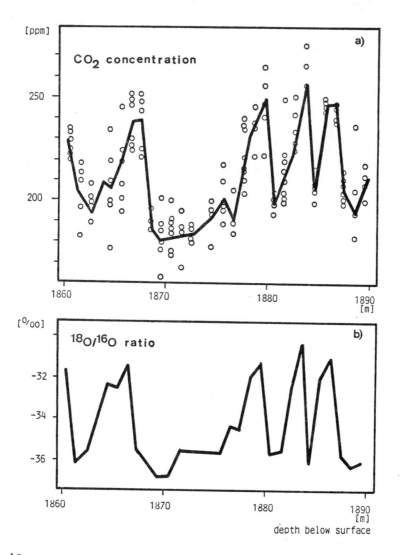

Fig. 10

CO_2-concentration and $^{18}O/^{16}O$ ratio at Dye 3 (southern Greenland, from Oeschger-Stauffer, 1983). The 30 m-layer is equivalent to an accumulation of about 10 000 years, 37-27 ka ago. Circles represent individual measurements, thick line time-averages.

two time-series with the equivalent values (between turning points) of the other series. Since the 3 m section has been accumulated during about 1000 years, we may assume that the height of the site remains nearly constant, i.e., that the difference of $\delta^{18}O$ represents true surface temperatures (Dansgaard, 1964). If these corresponding independent pairs are plotted (crosses) in a $CO_2/\delta^{18}O$-diagram (Fig. 11), a highly significant correlation coefficient of $r = 0.95$ is obtained. The lack of crosses near 240 ppm and $\delta^{18}O \sim -32$ indicates the abruptness of the climatic discontinuity. The linear regression equation yields

$$CO_2 \text{ (ppm)} = 617 + 11.7 \, \delta^{18}O \, (^o/oo).$$

Limiting our considerations to this 1000^a-period, representing the abrupt cooling between an interstadial and a stadial, we may assume that Dansgaard's (1964) linear relation between $\delta^{18}O$ of falling snow and surface temperature t_s is also representative for this period:

$$\delta^{18}O = 0.7 \, t_s - 13.6.$$

This allows us to combine these two empirical relations for the difference ΔCO_2

$$\Delta CO_2 = 11.7 \times 0.7 \, \Delta t_s \sim 8.2 \, \Delta t_s$$

$$\Delta t_s = \frac{66}{8.2} \sim 8^\circ C.$$

This value gives a first-order estimate of the local warming at southern Greenland; certainly the $\delta^{18}O$-value cannot be taken as representative for the globe in contrast to the CO_2-value. Under actual climatic conditions the temperature fluctuations in the Arctic are nearly three times as large as the simultaneous hemispheric changes; this should have been valid also 40 ka ago. For comparison, the same procedure - although less justified - has been used for the 10 ka series of our Fig. 10 (Oeschger-Stauffer, Fig. 7). This gives (with $r = 0.82$) the regression equation

$$CO_2 \text{ (ppm)} = 537 + 9.47 \, \delta^{18}O \, (^o/oo).$$

The hypothesis that the slope coefficients of both series (11.7 and 9.47) represent the same ensemble, should be rejected (at the 95% significance level). Thus it is not advisable to combine both results.

It could be conjectured that the amount of nutrients available in the shallow (and horizontally limited) circulation correlated with the equatorial undercurrent is limited. This limits the duration of the

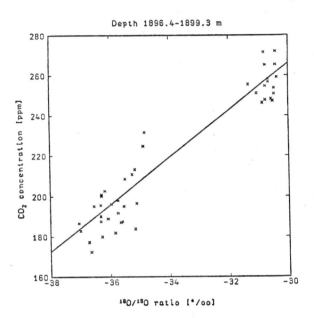

Fig. 11
Regression of $^{18}O/^{16}O$ values ($^o/oo$) and corresponding CO_2-concentrations in an ice-core from Greenland. The depth 1896.4-1899.3 m represents a time of ca. 1000 years. Data are taken from Oeschger and Stauffer (1983, Fig. 8).

prevailing oceanic flow pattern responsible for either decrease or increase of atmospheric CO_2. This has been suggested orally by Dr. W. Roether (Heidelberg) and could so be responsible for the apparently constant difference of 60-80 ppm CO_2 between the two modes (Fig. 10). In what way the quite complex abyssal circulation of the ocean accounts for a duration of a few millennia of each mode, remains a challenge.

This is a convincing example of an abrupt cooling under completely natural conditions, accompanied by a CO_2-drop of 60-70 ppm. The last examples have been experienced by late palaeolithic man in Europe during the Late Glacial and its transition to the Holocene: warming at the beginning of the Bölling-Alleröd warm epoch (ca. 12.8 ka BP), onset and end of the last cold epoch of the Younger Dryas (10.8 and 10.2 ka BP). In all three cases the summer temperature difference was in the order of $5^{\circ}C$ (Eicher, 1980; Oeschger et al., 1983, Fig. 2-4), apparently simultaneous with Greenland. The CO_2-content varied simultaneously by 60-80 ppm up and down (Oeschger-Stauffer, 1983, Fig. 6).

These observed facts are confirmed by many other data from ice cores, fossil lakes, ocean bottom cores (Schnitker, 1982) etc. He suggests a climatic feedback mechanism via the slow transport of bottom water originating in the subarctic Atlantic and flowing across the southern Indian Ocean towards the northern Pacific. The sea-surface temperature anomalies produced there - with a delay in the order of a millenium - may modulate via atmospheric flow patterns the circulation anomalies (and thus the intensity of the bottom water production) in the northernmost Atlantic. They resemble also the sequence of glacial advances and retreats in the Alps (Gamper-Suter, 1982), in northern Scandinavia (Karlén, 1976, 1979) and Alaska (Denton-Karlén, 1973). They suggest irregular, non-periodic (!) abrupt shifts between two climatic modes - "cold" versus "warm", both at different levels during glacials and interglacials - with a time-scale of a few thousand years, averaging at about 2500 years. The abrupt transitions between the two modes are distinctly shorter: a few examples suggest a typical transition time of 100 years. These features confirm, on this scale, Lorenz's suggestion (1970) of an almost-intransitive behavior of the climate system and resemble Saltzman's (1980, 1982) auto-regressive fluctuations of the climatic system. It should also be remembered that C.G. Rossby in his last paper (1959) suggested a time-scale of climatic fluctuations around 3000 years, if one takes the slow deep ocean circulation into account. However, these suggestions are not yet outlined in detail.

REFERENCES

Angell, J.K., 1981: Comparison of variations in atmospheric quantities with sea surface temperature variations in the equatorial eastern Pacific.
Mon. Wea. Rev., 109, 230-243.

Bacastow, R.B., and C.D. Keeling, 1981: Atmospheric CO_2 and the southern oscillation: effects associated with recent El Niño events. Reprints WMO/UNEP/ICSU Conf. on Analysis and Interpretation of CO_2 data, Bern, 109-112.

Baes jr., C.F., 1982: Effects of ocean chemistry and biology on atmospheric carbon dioxide. In: W.C. Clark (Ed.): Carbon Dioxide Review 1982, Oxford Univ. Press, 187-204.

Baes jr., C.F., 1983: The role of oceans in the carbon cycle. In: W. Bach et al. (Eds.): Carbon Dioxide-Current views and Developments in Energy/Climate Research. D. Reidel Publ. Comp. Dordrecht, 31-56.

Barnett, T.P., 1977: An attempt to verify some theories of El Niño.
J. Phys. Oceanogr., 7, 633-647.

Barnola, J.M., et al., 1983: Comparison of CO_2 measurements by two laboratories on air from bubbles in polar ice.
Nature, 302, 410-413.

Bartels, J., 1935: Zur Morphologie geophysikalischer Zeitfunktionen.
Sitz.ber. Preuss. Akad. Wiss. Berlin, Phys. math. Kl., 504-522.

Becker, C., 1982: Über die Zusammenhänge zwischen den Niederschlägen in Tropisch-Amerika und großräumigen Zirkulationen.
Diploma Thesis, Univ. Bonn; this volume, Part B.

Behrend, H., 1983: Telekonnektionen tropischer Niederschlagsanomalien.
Diploma Thesis, Univ. Bonn; this volume, Part A.

Berlage, H.P., 1957: Fluctuations of the general atmospheric circulation of more than one year, their nature and prognostic value.
Roy. Neth. Met. Inst., Meded. Verhandl., 69, 152 pp.

Berner, W., et al., 1980: Information of the CO_2 cycle from ice core studies.
Radiocarbon, 22, 227-235.

Bjerknes, J., 1966: A possible response of the atmospheric Hadley circulation and equatorial anomalies of ocean temperature.
Tellus, 18, 820-837.

Bjerknes, J., 1969: Atmospheric teleconnections from the equatorial Pacific.
Mon. Wea. Rev., 97, 163-172.

Broecker, W.S., 1982: Glacial to interglacial changes in ocean chemistry.
Progress in Oceanography, 11, 151-197.

Broecker, W.S., and T.-H. Peng, 1982: Tracers in the sea.
Lamont-Doherty Geological Observation, Columbia Univ. Palisades, N.Y.

Broecker, W.S., 1984: Carbon dioxide circulation through ocean and atmosphere.
Nature, 308, 602.

Brooks, C.E.P., and H.W. Braby, 1921: The clash of the trades in the Pacific.
Quart. Journ. Roy. Met. Soc., 47, 1-13.

Cane, M.A., 1983: Oceanographic events during El Niño.
Science, 222, 1189-1195.

Caviedes, C.N., 1973: Sêcas and El Niño: two simultaneous climatical hazards in South America.
Proc. Asso. Amer. Geogr., 5, 44-48.

Chu, P.Sh., 1983: Diagnostic studies of rainfall anomalies in northeast Brazil.
Mon. Wea. Rev., 111, 1655-1664.

Corneja-Garido, A.G., and P.H. Stone, 1977: On the heat balance of the Walker circulation.
J. Atmos. Sci., 34, 1155-1162.

Coy, L., 1979: An unusually westerly amplitude of the Quasi-Biannual Oscillation.
J. Atmos. Sci., 36, 174-176.

Dansgaard, W., 1964: Stable isotopes in precipitation.
Tellus, 16, 436-468.

Dansgaard, W., et al., 1982: A new Greenland deep ice core.
Science, 218, 1273-1277.

Delmas, R.J., et al., 1980: Polar ice evidence that atmospheric CO_2 30,000 yr. BP was 50% of present.
Nature, 284, 155-157.

Denton, G.H., and W. Karlén, 1973: Holocene climatic variations - their pattern and possible cause.
Quatern. Res., 3, 155-205.

Doberitz, R., H. Flohn and K. Schütte, 1967: Statistical investigations of the climatic anomalies of the equatorial Pacific.
Bonner Meteorol. Abh., 7.

Doberitz, R., 1968: Cross spectrum analysis of rainfall and sea temperature at the equatorial Pacific Ocean.
Bonner Meteorol. Abh., 8.

Doberitz, R., 1969: Cross-spectrum and filter analysis of monthly rainfall and wind data in the tropical Atlantic region.
Bonner Meteorol. Abh., 11, 53 pp.

Donguy, J.R., and C. Henin, 1978: Surface salinity fluctuations between 1956 and 1973 in the western south Pacific Ocean.
J. Phys. Oceanogr., 8, 1932-1934.

Donguy, J.R., and C. Henin, 1980: Climatic teleconnections in the western south Pacific with El Niño phenomenon.
J. Phys. Oceanogr., 10, 1952-1958.

Eicher, U., 1980: Mitt. Naturforsch. Ges. Bern, N.F., 65-80; cf. Quatern. Res., 15 (1981), 160-170.

Eickermann, W., and H. Flohn, 1962: Witterungszusammenhänge über dem äquatorialen Südatlantik.
Bonner Meteorol. Abh., 1, 62 pp.

Elliott, W.P., 1984: The pre-1958 atmospheric concentration of carbon dioxide.
EOS (Transact. Amer. Geophys. Un.), 65, 416-417.

Enfield, D.B., and J.S. Allen, 1980: On the structure and dynamics of monthly mean sea level anomalies along the Pacific coast of North and South America.
J. Phys. Oceanogr., 8, 557-578.

Fiadeiro, M.E., 1983: Physical-chemical processes in the open ocean. In: B. Bolin, R.B. Cook (Eds.) The Major Biochemical Cycles and their Interaction. J. Wiley & Sons, SCOPE Vol. 21, 461-476.

Fleer, H., 1975: Spektrum and Kreuspektrumanalyse von Niederschlagsreihen aus Indonesien, Australien und Westpazifik.
Diploma Thesis, Univ. Bonn.

Fleer, H., 1981: Large-scale tropical rainfall anomalies.
Bonner Meteorol. Abh., 26, 114 pp.

Flohn, H., and K.H. Hinkelmann, 1953: Äquatoriale Zirkulationsanomalien und ihre klimatische Auswirkung.
Ber. Dt. Wetterdienst US-Zone, 42, 114-121.

Flohn, H., 1957: Studien zur Dynamik der äquatorialen Atmosphäre.
Beitr. Phys. Atmos., 30, 18-46.

Flohn, H., 1966: Warum ist die Sahara trocken?
Z. f. Meteor., 17, 316-320.

Flohn, H., 1967: Bemerkungen zur Asymmetrie der atmosphärischen Zirkulation.
Ann. Meteor., N.F., 3, 76-80.

Flohn, H., 1971: Tropical circulation patterns.
Bonner Meteorol. Abh., 15.

Flohn, H., 1979: On time-scale and causes for abrupt climatic changes.
Quatern. Res., 12, 135-149.

Flohn, H., 1981: Klimaänderungen als Folge der CO_2-Zunahme?
Physik. Blätter 37, 184-190.

Flohn, H., 1982: Oceanic upwelling as a key for abrupt climatic change.
Journ. Meteor. Soc. Japan, 60, 268-273.

Flohn, H., 1983a: A climatic feedback mechanism involving oceanic upwelling, atmospheric CO_2 and water vapour.
In: A. Street-Perrott et al. (Eds.): Variations in the global water budgets. D. Reidel Publ. Comp., Dordrecht, 403-418.

Flohn, H., 1983b: Actual palaeoclimatic problems from a climatologist's viewpoint.
In: A. Ghazi (Ed.): Palaeoclimatic Research and Models. D. Reidel Publ. Comp., Dordrecht, 17-29.

Flohn, H., 1984: A possible mechanism of abrupt climatic changes.
In: N.A. Mörner, W. Karlén (Eds.): Climatic Changes on a Yearly to Millennial Basis. D. Reidel Publ. Comp., Dordrecht, 521-531.

Frank, W.M., 1983: The structure and energetics of the east Atlantic intertropical convergence zone.
J. Atmos. Sci., 40, 1916-1929.

Gamper, M., and J. Suter, 1982: Postglaziale Klimageschichte der Schweizer Alpen.
Geographica Helvetica, 37, 105-114.

Hastenrath, S., and L. Heller, 1977: Dynamics of climatic hazards in Northeast Brazil.
Quart. J. Roy. Met. Soc., 103, 77-92.

Hastenrath, S., and E.B. Kaczmarczyk, 1981: On spectra and coherence of tropical climate anomalies.
Tellus, 33, 453-462.

Helbig, M., 1976: Korrelations- und Spektrumanalyse von zusätzlichen Niederschlagsreihen aus Afrika und Reduktion der Niederschlagsdaten durch Gebietsmittelung.
Appendix of the Diploma Thesis, Univ. Bonn.

Henning, D., and H. Flohn, 1980: Some aspects of evaporation and sensible heat flux in the tropical Atlantic.
Contrib. Atmosph. Phys., 53, 430-441.

Ichiye, T., and J.R. Petersen, 1963: The anomalous rainfall of the 1957-58 winter in the equatorial central Pacific arid area.
J. Meteor. Soc. Japan, 41, 172-182.

Julian, P.R., and R.M. Chervin, 1978: A study of the Southern Oscillation and Walker circulation phenomenon.
Mon. Wea. Rev., 106, 1433-1451.

Karlén, W., 1979: Glacier variations in the Svartisen area, Northern Norway. Geografiska Annaler 61 A, 11-28; cf. also l.c. 58 A, 1976, 1-34.

Keeling, C.D., et al., 1984: Seasonal, latitudinal and secular variations on the abundance and isotope ratios of atmospheric CO_2, Part 2.
J. Geophys. Res., 89, 4615-4628.

Keen, R.A., 1982: The role of cross-equatorial tropical cyclone pairs in the southern oscillation.
Mon. Wea. Rev., 110, 1405-1416.

Kellogg, W.W., 1983: Feedback mechanisms in the climatic system affecting future levels of carbon dioxide.
J. Geophys. Res., 88, 1263-1269.

Keshavamurty, R.N., 1982: Response of the atmosphere to sea surface temperature anomalies over the equatorial Pacific and the teleconnections of the Southern Oscillation.
J. Atmos. Sci., 39, 1241-1259.

Knoch, K., 1927: Große Anomalien des Niederschlags in der Äquatorzone des Pazifischen Ozeans.
Ann. Hydrogr. Mar. Meteor., 55, 361-367.

Krueger, A.F., and J.S. Winston, 1974: A comparison of the flow over the tropics during two contrasting circulation regimes.
J. Atmos. Sci., 31, 358-370.

Krueger, A.F., and J.S. Winston, 1975: Large-scale circulation anomalies during 1971-72.
Mon. Wea. Rev., 103, 465-473.

Lenz, J., 1981: Produktionsbiologische Bedeutung von Auftriebsvorgängen im Meer.
Naturwiss. Rdsch., 34, 405-413.

Lettau, H., 1956: Theoretical notes on the dynamics of the equatorial atmosphere.
Beitr. Phys. Atmos., 29, 107-122.

Lorenz, E., 1970: Climatic change as a mathematical problem.
J. Appl. Meteor., 9, 325-329.

Lorius, C., and D. Raynaud, 1983: Record of past atmospheric CO_2 from tree-ring and ice-core studies.
In: W. Bach et al. (Eds.): Carbon-Dioxide - Current Views and Developments in Energy/Climate Research. D. Reidel Publ. Comp., 145-177.

Luther, D.S., and D.E. Harrison, 1984: Observing long-period fluctuations of surface winds in the Tropical Pacific.
Mon. Wea. Rev., 113, 285-302; cf. also Science, 222, 1983, 327-330; 224, 1984, 1099-1102.

Markham, C.G., 1974: Apparent periodicities in rainfall at Fortalezza, Ceará, Brazil.
J. Appl. Meteor., 13, 176-179.

Meissner, B.N., 1976: A study of Hawaiian and Line Islands rainfall.
UH-MET 76-04, Univ. of Hawaii, 83 pp.

Monthly Climatic Data for the World, 1972-1980, Asheville, N.C.

Namias, J., and D.R. Cayan, 1981: Large-scale air-sea interactions and short-period climatic fluctuations.
Science, 214, 869-876.

Newell, R.E., R. Navato and J. Hsiung, 1978: Long-term global sea surface temperature fluctuations and their possible influence on atmospheric CO_2 concentration.
Pure and Appl. Geophys., 116, 351-371.

Newell, R.E., and J. Hsiung, 1984: Sea surface temperature, atmospheric CO_2 and the global energy budget: some comparisons between the past and present.
In: N.A. Mörner, W. Karlén (Eds.): Climatic Changes on a Yearly to Millennial Basis. D. Reidel Publ. Comp., Dordrecht, 533-561.

Nicholls, N., 1981: Air-sea interaction and the possibility of long-range weather prediction in the Indonesian Archipelago.
Mon. Wea. Rev., 109, 2435-2443.

Nicholson, Sh.E., 1983: Sub-Saharan rainfall in the years 1976-80: evidence of continued drought.
Mon. Wea. Rev., 111, 1646-1654.

Oeschger, H., et al., 1983: Late-glacial climate history from ice cores.
In: A. Ghazi (Ed.): Palaeoclimatic Research and Models. D. Reidel Publ. Comp., Dordrecht, 95-107.

Oeschger, H., and B. Stauffer, 1983: Review of the history of the atmospheric CO_2 recorded in ice cores.
Paper presented at the ORNL Life Science Symposium on the Global Carbon Cycle, Knoxville, Oct. 31 - Nov. 2.

Oort, A., 1983: Global atmospheric circulation statistics 1958-73.
U.S. Department of Commerce, NOAA Professional Paper 15.

Pan, Y.H., and A.H. Oort, 1983: Global climate variation connected with sea surface temperature anomalies in the eastern equatorial Pacific Ocean for the 1958-73 period.
Mon. Wea. Rev., 111, 1244-1253.

Philander, S., 1981: The response of equatorial oceans to a relaxation of the trade winds.
J. Phys. Oceanogr., 11, 176-189.

Philander, S., 1983: El Niño/Southern Oscillation phenomena.
Nature, 302, 295-301.

Quinn, W.H., and W.V. Burt, 1972: Use of the Southern Oscillation in weather prediction.
J. Appl. Meteor., 11, 616-628.

Raatz, W., 1977: Räumliche Korrelationen und Variabilität der Niederschläge auf Sri Lanka und Südindien (Nilgiris).
Diploma Thesis, Univ. Bonn.

Ramage, C.S., 1968: Role of a tropical 'maritime continent' in the atmospheric circulation.
Mon. Wea. Rev., 96, 365-370.

Ramage, C.S., 1975: Preliminary discussion of the meteorology of the 1972-73 El Niño.
Bull. Amer. Meteor. Soc., 56, 234-242.

Ramage, C.S., 1977: Sea surface temperature and local weather.
Mon. Wea. Rev., 105, 540-544.

Ramage, C.S., S.J.S. Khalsa and B.N. Meisner, 1981: The central Pacific near-equatorial convergence zone.
J. Geophys. Res., 86, 6580-6598.

Ramanathan, V., 1981: The role of ocean-atmosphere interactions in the CO_2 climate problem.
J. Atmos. Sci., 38, 918-930.

Rasmussen, E.M., and T.H. Carpenter, 1982: Variations in tropical sea surface temperature and surface wind fields associated with the Southern Oscillation/El Niño.
Mon. Wea. Rev., 110, 354-384.

Rasmussen, E.M., and J.M. Wallace, 1983: Meteorological aspects of the El Niño/Southern Oscillation.
Science, 222, 1195-1202.

Reiter, E.R., 1978: The interannual variability of the ocean-atmosphere system.
J. Atmos. Sci., 35, 349-370.

Reiter, E.R., 1983: Teleconnections with tropical precipitation surges.
J. Atmos. Sci., 40, 1631-1647.

Rossby, C.G., 1959: Current problems in Meteorology.
In: B. Bolin (Ed.): The Atmosphere and Sea in Motion. Rossby Memorial Volume, 9-50.

Rowntree, P.R., 1972: The influence of tropical east Pacific Ocean temperature on the atmosphere.
Quart. J. Roy. Met. Soc., 98, 290-321.

Sadler, J.C., 1969: Average cloudiness in the tropics from satellite observations.
International Indian Ocean Expedition Meteor. Monogr., No. 2, East-West Center Press.

Saltzman, B., and R.E. Moritz, 1980: A time-dependent climatic feedback system involving sea-ice extent, ocean temperature and CO_2.
Tellus, 32, 93-118; cf. l.c. 34, 1982, 97-112.

Schnell, R.C., J.M. Harris and J.A. Schroeder, 1981: A relationship between Pacific Ocean temperatures and atmospheric carbon dioxide concentrations at Barrow and Mauna Loa.
Reprints WMO/UNEP/ICSU Conf. on Analysis and Interpretation of CO_2 data, Bern, 155-162.

Schnitker, D., 1982: Climatic variability and deep ocean circulation: Evidence from the North Atlantic.
Palaeogeogr., Palaeoclim. Palaeoecol., 40, 213-234.

Schweitzer, B., 1978: Gebietsmittel tropischer Niederschläge.
Appendix of the Diploma Thesis, Univ. Bonn.

Seckel, G.R., and M.Y.Y. Young, 1971: Harmonic functions for sea surface temperatures and salinities, Koko Head, Oahu, 1956-69, and sea surface temperature, Christmas Island, 1954-69.
Fish. Bull., 69, 181-214.

Seckel, G.R., and M.Y. Young, 1977: Koko Head, Oahu sea surface temperatures and salinity, 1956-73, and Christmas Island sea surface temperatures, 1954-1973.
Fish. Bull., 75, 767-787.

Shukla, J., 1975: Effect of Arabian sea surface temperature anomaly on Indian summer monsoon: a numerical experiment with the GFDL-model.
J. Atmos. Sci., 32, 503-511.

Siegenthaler, U., and Th. Wenk, 1984: Rapid atmospheric CO_2 variations and ocean circulation.
Nature, 308, 624-626.

Sumi, A., and T. Murakami, 1981: Large-scale aspects of the 1978-79 winter circulation over the greater WMONEX region.
J. Meteor. Soc. Japan, 59, 625-645.

Taylor, R.C., 1973: An atlas of Pacific island rainfall.
Hawaii Inst. Geophys. Data Rep. No. 25, HIG-73-9.

Trempel, U., 1978: Eine klimatologische Auswertung der meteorologischen Beobachtungen deutscher Handelsschiffe vor der Westküste Südamerikas aus dem Zeitraum 1869-1970.
Diploma Thesis, Univ. Bonn.

Trenberth, K.E., 1976: Spatial and temporal variations of the Southern Oscillation.
Quart. J. Roy. Met. Soc., 102, 639-653.

Troup, A.J., 1965: The Southern Oscillation.
Quart. J. Roy. Met. Soc., 91, 490-506.

Vitousek, M.J., 1979: Line Islands Monitoring.
NORPAX Quarterly Report, 1. Jan. - 31. March 1979, 40-43.

Weare, B.C., 1984: Interannual moisture variations near the surface of the tropical Pacific Ocean.
Quart. J. Roy. Met. Soc., 110, 489-504.

Weber, K.-H., 1981: Abschätzung des Energieaustausches an der Meeresoberfläche im Arabischen und Roten Meer.
Diploma Thesis, Univ. Bonn.

Wei, M.Y., D.R. Johnson, R.D. Townsend, 1983: Seasonal distribution of diabatic heating during the First GARP Global Experiment.
Tellus, 35A, 241-255.

Weichert, G., 1980: Chemical changes and primary production in upwelled water off Northwest Africa.
Dt. Hydrogr. Zeitschr., 33, 192-198.

Wells, N.C., 1979: The effect of a tropical sea surface temperature anomaly in a coupled ocean-atmosphere model.
J. Geophys. Res., 84, 4985-4997.

Woodley, W.L., C.G. Griffith, J.S. Griffin and S.C. Stromatt, 1980: The inference of GATE convective rainfall from SMS-1 imagery.
J. Appl. Meteor., 19, 388-408.

Wooster, W.S., and O. Guillan, 1974: Characteristics of El Niño 1972.
J. Mar. Res., 32, 387-404.

Wright, P.B., 1975: An index of the Southern Oscillation.
Climatic Res. Unit, CRU RP4, Univ. East Anglia, Norwich, England.

Wright, P.B., 1977: The Southern Oscillation - patterns and mechanisms of the teleconnections and the persistence.
Hawaii Inst. Geophys., Rep. HIG 77-13.

Wright, P.B., 1979: Persistence of rainfall anomalies in the Central Pacific.
Nature, 277, 371-374.

Wyrtki, K., 1973: Teleconnections in the equatorial Pacific Ocean.
Science, 180, 66-68.

Wyrtki, K., 1975: El Niño - the dynamic response of the equatorial Pacific Ocean to atmospheric forcing.
J. Phys. Oceanogr., 5, 572-584.

Wyrtki, K., 1977: Sea level during the 1972 El Niño.
J. Phys. Oceanogr., 7, 779-787.

Wyrtki, K., 1979: The response of sea surface topography to the 1976 El Niño.
J. Phys. Oceanogr., 9, 1223-1231.

Wyrtki, K., 1981: En estimate of equatorial upwelling in the Pacific.
J. Phys. Oceanogr., 11, 1205-1214.

Wyrtki, K., and G. Eldin, 1982: Equatorial upwelling events in the Central Pacific.
J. Phys. Oceanogr., 12, 984-988.

Wyrtki, K., and B. Kilonsky, 1984: Mean water and current structure during the Hawaiian-to-Tahiti Shuttle experiment.
J. Phys. Oceanogr., 14, 242-254.

Yoshida, K., and H.-L. Mao, 1957: A theory of upwelling of large horizontal extent.
J. Mar. Res., 16, 40-54.

BONNER METEOROLOGISCHE ABHANDLUNGEN

Herausgegeben vom Meteorologischen Institut der Universität Bonn durch Prof.Dr. H. FLOHN (Heft 1-25), ab Heft 26 durch Prof.Dr. M. HANTEL.

Heft 1: *W. Eickermann und H. Flohn:* Witterungszusammenhänge über dem äquatorialen Südatlantik. 1962, 65 S. vergr.

Heft 2: *Hermann Flohn:* Klimaschwankungen und großräumige Klimabeeinflussung. 1963, 61 S. vergr.

Heft 3: *Masatoshi M. Yoshino:* Rainfall, Frontal Zones and Jet Streams in Early Summer over East Asia. 1963, 126 S. vergr.

Heft 4: *Hermann Flohn:* Investigations on the Tropical Easterly Jet. 1964, 83 S. DM 10,—

Heft 5: *Hermann Flohn:* Studies on the Meteorology of Tropical Africa. 1965, 57 S. vergr.

Heft 6: *H. Flohn, D. Henning, H.C. Korff:* Studies on the Water-Vapor Transport over Northern Africa. 1965, 63 S. vergr.

Heft 7: *R. Doberitz, H. Flohn, K. Schütte:* Statistical Investigations of the Climatic Anomalies of the Equatorial Pacific. 1967, 78 S. DM 10,—

Heft 8: *Rolf Doberitz:* Cross-Spectrum Analysis of Rainfall and Sea Temperature at the Equatorial Pacific Ocean. — A Contribution to the El-Niño-Phenomenon. 1968, 62 S. DM 10,—

Heft 9: *Karin Schütte:* Untersuchungen zur Meteorologie und Klimatologie des El-Niño-Phänomens in Ecuador und Nordperu. 1968, 152 S. DM 16,—

Heft 10: *J.-O. Strüning, H. Flohn:* Investigations on the Atmospheric Circulation above Africa. 1969, 56 S. DM 10,—

Heft 11: *Rolf Doberitz:* Cross-Spectrum and Filter Analysis of Monthly Rainfall and Wind Data in the Tropical Atlantic Region. 1969, 53 S. DM 10,—

Heft 12: *Michael Agi:* Globale Untersuchungen über die räumliche Verteilung der kinetischen Energie in der Atmosphäre. 1970, 78 S. DM 10,—

Heft 13: *Jens-Ole Strüning:* Untersuchungen zur Divergenz des Wasserdampftransportes in Nordwestdeutschland. 1970, 61 S. DM 10,—

Heft 14: *H. Flohn, M. Hantel, E. Ruprecht:* Investigations on the Indian Monsoon Climate. 1970, 100 S. DM 10,—

Heft 15: *Hermann Flohn:* Tropical Circulation Pattern. 1971, 55 S. vergr.

Heft 16: *Frank Schmidt:* Entwurf eines Modells zur allgemeinen Zirkulation der Atmosphäre und Simulation klimatologischer Strukturen. 1971, 68 S. DM 10,—

Heft 17: Klimatologische Forschung. Festschrift für Hermann Flohn zur Vollendung des 60. Lebensjahres. Climatological Research. The Hermann Flohn 60th Anniversary Vol. Hrsg. von *K. Fraedrich, M. Hantel, H. Clausen Korff, E. Ruprecht.* 1974, XIV, 610 S., 46 Beiträge (22 in deutscher und 24 in englischer Sprache), 54 Tabellen, 262 Abb.   DM 118,—

Heft 18: *Michael Hantel:* Ein vertikal-integriertes Modell der Passatschicht. 1973, 60 S.   DM 18,—

Heft 19: *Hermann Flohn* (Sonderheft): Globale Energiebilanz und Klimaschwankungen. Erhältlich nur über: Meteorologisches Institut der Universität Bonn. D 53 Bonn. Auf dem Hügel 20.

Heft 20: *Klaus Fraedrich:* Energetik synoptischer Störungen mit kaltem Kern in der tropischen Atmosphäre. 1974, 67 S.   DM 22,—

Heft 21: *Hermann Flohn* (Sonderheft): Tropische Zirkulationsformen im Lichte der Satellitenaufnahmen. Erhältlich nur über: Meteorologisches Institut der Universität Bonn. D 53 Bonn, Auf dem Hügel 20.

Heft 22: *Wolfgang Peyinghaus:* Eine numerische Berechnung der Strahlungsbilanz und die Strahlungserwärmung der Atmosphäre im Meridional-Vertikalschnitt. 1974, 64 S.   DM 22,—

Heft 23: *M. Wagner, E. Ruprecht:* Materialien zur Entwicklung des indischen Sommermonsuns. 1975, 104 S.   DM 26,—

Heft 24: *Wilfried Bach:* Changes in the Composition of the Atmosphere and their Impact upon climatic Variability — an Overview. 1976, 51 S.   DM 22,—

Heft 25: *K. Fraedrich, H. Nitsche, B. Rudolf, W. Thommes, W. Wergen:* Konvektion und Diffusion über einem Kühlturm — Entwurf eines Modells —. 1978, 73 S.   DM 26,—

Heft 26: *H. Fleer:* Large-Scale Tropical Rainfall Anomalies. 1981, 124 S.   DM 36,—

Heft 27: *Jörg M. Hacker:* Der Massen- und Energiehaushalt der Nordhemisphäre. 1981, 102 S.   DM 30,—

Heft 28: *Michael Memmesheimer:* Klimatheorie mit Spektralmodellen niedrigster Ordnung. 1982, 126 S. + X.   DM 36,—

Heft 29: *Michael Hantel and Sabine Haase:* Mass Consistent Heat Budget of the Zonal Atmosphere. 1983, 84 S. + X.   DM 30,—

Heft 30: *Dieter Klaes:* A Cloud/Radiation Model for Synoptic Energy Budgets. 1984, 114 S. + X.   DM 36,—

In Kommission bei Ferd. Dümmlers Verlag, Kaiserstr. 31—37, 5300 Bonn 1